Professional Development of Chemistry Teachers
Theory and Practice

Advances in Chemistry Education Series

Editor-in-chief:
Keith S. Taber, *University of Cambridge, UK*

Series editors:
Avi Hofstein, *Weizmann Institute of Science, Israel*
Vicente Talanquer, *University of Arizona, USA*
David Treagust, *Curtin University, Australia*

Titles in the series:
1: Professional Development of Chemistry Teachers: Theory and Practice

How to obtain future titles on publication:
A standing order plan is available for this series. A standing order will bring delivery of each new volume immediately on publication.

For further information please contact:
Book Sales Department, Royal Society of Chemistry, Thomas Graham House, Science Park, Milton Road, Cambridge, CB4 0WF, UK
Telephone: +44 (0)1223 420066, Fax: +44 (0)1223 420247,
Email: booksales@rsc.org
Visit our website at www.rsc.org/books

Professional Development of Chemistry Teachers
Theory and Practice

Rachel Mamlok-Naaman
Weizmann Institute of Science, Israel
Email: rachel.mamlok@weizmann.ac.il

Ingo Eilks
University of Bremen, Germany
Email: ingo.eilks@uni-bremen.de

George Bodner
Purdue University, USA
Email: gmbodner@purdue.edu

and

Avi Hofstein
Weizmann Institute of Science, Israel
Email: avi.hofstein@weizmann.ac.il

ROYAL SOCIETY
OF **CHEMISTRY**

Advances in Chemistry Education Series No. 1

Hardback ISBN: 978-1-78262-706-7
PDF ISBN: 978-1-78801-340-6
EPUB ISBN: 978-1-78801-456-4
Print ISSN: 2056-9335
Electronic ISSN: 2056-9343
Paperback ISBN: 978-1-83916-742-3

A catalogue record for this book is available from the British Library

The Royal Society of Chemistry is a charity, registered in England and Wales, Number 207890, and a company incorporated in England by Royal Charter (Registered No. RC000524), registered office: Burlington House, Piccadilly, London W1J 0BA, UK, Telephone: +44 (0) 207 4378 6556.

For further information see our web site at www.rsc.org

Printed in the United Kingdom by CPI Group (UK) Ltd, Croydon, CR0 4YY, UK

Acknowledgements

We would like to thank Professor Anat Yarden and the Ilse Katz Foundation, the Weizmann Institute of Science, for supporting the editing of the book. We are also grateful to Camille Weinstein for helping us edit the book.

Advances in Chemistry Education Series No. 1
Professional Development of Chemistry Teachers: Theory and Practice
By Rachel Mamlok-Naaman, Ingo Eilks, George Bodner and Avi Hofstein
© Rachel Mamlok-Naaman, Ingo Eilks, George Bodner and Avi Hofstein 2022
Published by the Royal Society of Chemistry, www.rsc.org

About the Authors

Dr Rachel Mamlok-Naaman studied chemistry and chemistry education. She is employed in the chemistry group at the Department of Science Teaching, the Weizmann Institute of Science, where she served both as the head of the National Center for Chemistry Teachers until December 2015, and as the coordinator of the chemistry group at the Department of Science Teaching (until June 2016). Thus, her publications focus on topics related to students' learning and to teachers' professional development. For her work on chemistry teachers' professional development in Israel, she received the 2016 Maxine Singer Prize for outstanding scientists at the Weizmann Institute. Since 1982, she uses in her work the experience that she gained as a chemistry teacher, teaching high-school students for 26 years, and as a regional consultant for chemistry teachers. She is the coordinator of chemistry teachers' programs in the framework of the Rothschild-Weizmann MSc program for science teachers, and of projects in the framework of the European Union (the FP7 programs) in Israel: PARSEL (which ended in October 2010); PROFILES (which ended in July 2013), TEMI (which ended in July 2015), and an external evaluator of ARTIST, a project in the framework of Erasmus+. In addition: (i) an ACS senior member, and representative of the Israel Chemical Society to the various international organizations of chemistry education, including being a titular member of IUPAC CCE and the secretary of EuCheMS DivCED; (ii) the ICS contact person for all matters of chemistry education and relevant activities; and (iii) a member of editorial and advisory boards for journals and organizations of science education. Rachel's main focus was, and is, professional development of chemistry teachers (in-service and pre-service), in which she used her many years of experience as a high-school chemistry teacher. However, her contributions to chemistry education research and her publications are in several areas, which are integrated into each other, and guide her in conducting

Advances in Chemistry Education Series No. 1
Professional Development of Chemistry Teachers: Theory and Practice
By Rachel Mamlok-Naaman, Ingo Eilks, George Bodner and Avi Hofstein
© Rachel Mamlok-Naaman, Ingo Eilks, George Bodner and Avi Hofstein 2022
Published by the Royal Society of Chemistry, www.rsc.org

professional development courses for chemistry teachers, and in supporting individual teachers in their work: (i) development, implementation, and evaluation of new curricular materials, (ii) research on students' perceptions of chemistry concepts (especially bonding and structure), (iii) inquiry-type skills and activities, including teaching and learning in different cultures (*e.g.*, argumentation, asking questions, hypothesizing), (iv) relevance in chemistry education, (v) the nature of science, and (vi) education for sustainable development (ESD), for which she has been selected as a 2018 Awardee for the ACS-CEI Award for Incorporation of Sustainability into the Chemistry Curriculum.

Prof. Dr Ingo Eilks FRSC studied chemistry, mathematics, education, and philosophy in a program for prospective lower and upper secondary school teachers at the University of Oldenburg, Germany. He holds both a PhD and a habilitation in chemistry education and taught both chemistry and mathematics in different high schools for several years. Prof. Ingo Eilks has been involved in the pre-service education of chemistry teachers at various German universities since 1996. Since 2004, he has been a full professor in chemistry education and head of the chemistry education research group within the Department of Biology and Chemistry at the University of Bremen, Germany. For more than 20 years, Prof. Ingo Eilks held numerous courses for pre-service and in-service professional development of chemistry teachers in many different countries. The main focus of his professional development courses has been, among other topics, new ways toward the particulate nature of matter, cooperative learning in chemistry classes, use of modern information and communication technology in chemistry education, and societal-oriented and socio-scientific issues-based science education. He was an active part of a whole set of transnational projects on the professional development of science teachers funded by the European Union, namely the PROFILES, SALiS, TEMI and ARTIST projects. He is one of the editors (jointly with Prof. Avi Hofstein) of the book *Teaching Chemistry – A Study Book*, one of the very few international textbooks for the education and professional development of chemistry teachers. The group of Prof. Ingo Eilks is one of the most research active and internationally visible groups in chemistry education from Germany. The group is, however, also very active in professional development in the local and regional context around the University of Bremen. Two examples are: The group is one of two partners in a chemistry teacher in-service professional development center (*Chemielehrerfortbildungszentrum*) that is funded for about 20 years now by the German chemical society (GDCh). The group is also accompanying schools and teachers in different projects of action research. One outstanding project is a co-operation with a group of teachers that is now active in its 19th year and that is described in Chapter 5 of this book. The teaching, curriculum development and research of Prof. Ingo Eilks received numerous awards. Among the awards, one can find the University of Bremen teaching award (*Berninghausen Award*), the *STEM of Tomorrow*

School Award, three-times awards within the *United Nations Decade of Education for Sustainable Development (DESD)*, or recently the *Award for Outstanding Contributions to the Incorporation of Sustainability into Chemical Education* by the *American Chemical Society*.

Prof. Dr George M. Bodner FRSC, FACS started his undergraduate career as a history/philosophy major at the State University of New York at Buffalo. In his first year as an undergraduate he found that chemistry was "fun" and changed his major, graduating with a BS in chemistry in 1969. He began his career more than 50 years ago, doing synthetic inorganic chemistry. He soon learned that he was not good at synthetic chemistry, but fell in love with the potential of nuclear magnetic resonance (NMR) spectroscopy. He received his PhD in inorganic and organic chemistry from Indiana University in 1972 using Fourier Transform (FT) NMR to study organometallic compounds. He then took a position as a visiting assistant professor at the University of Illinois at Urbana-Champaign, where he taught general chemistry and biochemistry. The first step in the transition from traditional chemistry research to chemistry/science education research occurred in his first year at UIUC, while teaching general chemistry. In spite of what appeared to be well-crafted and well-delivered lectures on topics such as molarity, he found that no more than two-thirds of the bright, hard-working science and engineering majors in his course could successfully solve "simple" molarity problems. Although some of his publications on the use of FT NMR are still being cited today, he followed the advice of a senior colleague who argued that there is a difference between research that *could be done* and research that *should be done*. He therefore left Illinois in 1975 to take a job as two-thirds of the chemistry faculty at Stephens College, where he taught general, organic, and biochemistry and set the foundation for his future work on discipline-based educational research. The next step toward a career doing chemical education research occurred in 1977, when he took a faculty position at Purdue University, where he is now the Arthur E. Kelly Distinguished Professor of Chemistry, Education and Engineering. In 1981, he helped create the first graduate program in chemical education. In 2005, he was a member of the group that created the first School of Engineering Education at Purdue. The Division of Chemical Education at Purdue has graduated 100 PhDs and the School of Engineering Education has grown from four to more than 20 faculty. He is a Fellow of both the American Chemical Society and the Royal Society of Chemistry, a past chair of the ACS Division of Chemical Education, and recently completed a six-year term on the ACS Board of Directors. He has received ten awards at the department, college or university level for excellence in teaching. In 2013, he received the most prestigious award given to Purdue faculty, the *Morrill Award for Outstanding Achievement* given to faculty who have achieved a balance among excellence in teaching, research and service. This award was created in 2012 to celebrate the sesquicentennial of the

Morrill Act of 1862 that led to the creation of "land grant" universities, such as Purdue and UIUC. In 2003, he received the *George C. Pimentel Award in Chemical Education* and the *Nyholm Prize for Education* and was chosen to receive the *ACS Award for Achievement in Research for the Teaching and Learning of Chemistry*. He has been an associate editor of the *Journal of Research in Science Teaching, Chemistry Education Research and Practice*, and the *Journal of Science Teacher Education*. When he first applied discipline-based educational research to "advanced" undergraduate courses, such as the second-year course in organic chemistry, editors reacted to papers he submitted by arguing that "our readers would not understand the chemistry". Eventually he was able to convince some of them that the problem was that the editor did not understand the chemistry. He has published more than 150 papers, eight editions of general chemistry textbooks, and his group has presented more than 500 papers at technical conferences. His research has studied the challenges of teaching and learning chemistry from the introductory general chemistry course taken by science and engineering majors to the last stages of the preparation of graduate students working toward a PhD in chemistry. At the moment, his work is studying the emerging field of biochemistry education and continuing a longstanding commitment to both research and innovation oriented toward overcoming the barriers to success in the first- and second-year college chemistry courses for students who are blind or with low vision.

Prof. Dr Avi Hofstein holds a BSc in Chemistry from the Hebrew University in Jerusalem, an MA in Education from Tel-Aviv University, and a PhD in Science Education (Chemistry) from the Weizmann Institute of Science. For more than 15 years, Prof. Hofstein taught chemistry in two high schools in Israel. He served as head of the chemistry group, head of the national Center for Chemistry teachers, and head of the Department of Science Teaching of in the Weizmann Institute of Science in Israel. In recent years he also serves as the head of the research authority of the Academic Arabic College in Haifa. For almost 50 years, Prof. Hofstein has been involved in all facets of the curricular process in chemistry, including: Development, implementation, evaluation, and research. He has conducted research in the following areas: The science laboratory, classroom, outdoor and laboratory learning environments, learning difficulties, and students' misconceptions in chemistry, and attitudinal and motivational studies. He published to gather with his collegues and students more than 120 papers and book chapters. He was involved in three EU projects on pedagogy and professional development of science teachers. Recently he edited jointly with Professor Ingo Eilks two books related to chemistry teaching and learning titled: *Teaching Chemistry – A Study Book* and a book on *Relevant Chemistry Education*. In 2014 he was awarded the ACS-CEI, in 2016 the ACS award for his contribution to research in chemistry education, and in 2017 he was awarded the NARST DCRA (Distinguished Contribution for Research Award).

Contents

Advances in Chemistry Education Series No. 1
Professional Development of Chemistry Teachers: Theory and Practice
By Rachel Mamlok-Naaman, Ingo Eilks, George Bodner and Avi Hofstein
© Rachel Mamlok-Naaman, Ingo Eilks, George Bodner and Avi Hofstein 2022
Published by the Royal Society of Chemistry, www.rsc.org

CHAPTER 1

Introduction – Issues Related to the Professional Development of Chemistry Teachers

Professional development should be a continuing aspect of teachers' careers. It starts when teachers first enroll in pre-service programs and continues until they retire. There are many different models for initial preparation and continuing professional development for chemistry teachers in different countries around the world. This book was written by four authors from Israel, Germany and the USA. Many of the models, ideas, and activities that are presented in the book are based on the authors' personal involvement and research over a long period of time. As an introduction, this chapter discusses the different approaches to pre-service chemistry teacher education, consequences for continuous professional development, and the intentions of this book.

1.1 The Fields of Chemistry-Teaching Practices

Any discussion of the different approaches to both pre-service and in-service professional development of chemistry teachers should start with a look at the general differences in educational systems worldwide that impact the fields of practice for chemistry-teaching professionals.

There is general agreement in most countries that science at the primary-school level (mainly 6 to 10 years of age) should be taught using an integrated approach. This approach can focus primarily on science itself

Advances in Chemistry Education Series No. 1
Professional Development of Chemistry Teachers: Theory and Practice
By Rachel Mamlok-Naaman, Ingo Eilks, George Bodner and Avi Hofstein
© Rachel Mamlok-Naaman, Ingo Eilks, George Bodner and Avi Hofstein 2022
Published by the Royal Society of Chemistry, www.rsc.org

(integrating topics from biology, chemistry, and physics), or on science in one subject combined with other domains, such as history, geography, or technology. There also seems to be some consensus that chemistry at the upper secondary or high-school level (ages 15 or 16 to 18 years) should be taught as a standalone subject in its own right. Unfortunately, in many countries, upper secondary school chemistry is not required for all students, or is only compulsory for one year.

The largest differences in the way chemistry is introduced into the educational system can be found at what is known as the middle school, lower secondary school, or junior high-school level. This level usually covers students who are 10 to 14, 15 or 16 years of age. In some countries, science at this level is taught as an integrated science subject combining aspects of chemistry, biology, physics, and geoscience. In other countries, science is taught as largely independent subjects (chemistry, biology, and physics), quite frequently with biology or earth sciences being taught before chemistry starts (Figure 1.1). Sometimes the split in coverage of individual science subjects occurs somewhere midway through lower secondary school. In Germany, for example, there are schools in which the separation into individual science subjects begins in grades 5, 7, or 9, or at the start of upper secondary science education.

There is no clear evidence to show whether it is more effective to teach chemistry in middle school or at the lower secondary level as a subject in its own right, or integrated with the other domains of science. An advantage to

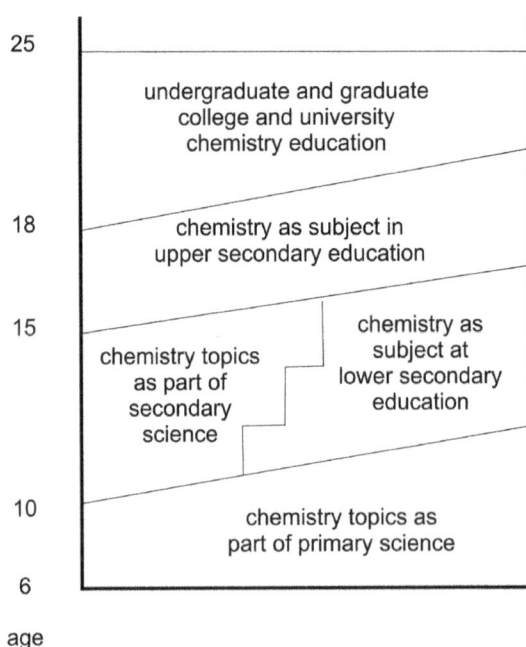

Figure 1.1   Domains of teaching chemistry.

chemistry as a standalone subject might be a greater concentration on the content matter and inner structure of chemistry, whereas an integrated approach might facilitate a broader view of chemistry, including technological applications and environmental or societal ramifications. As an independent subject, it might be easier to focus on the specific characteristics and nature of chemistry, whereas the integrated approach could be a better way to show what the different science domains have in common and how they are related.

For pre-service teacher education and continuous professional development, neither approach creates a structural problem as long as they are operated consistently throughout the educational system. For educational systems that mix the two approaches, both teachers and providers of continuous professional development regularly face difficulties. This is the case in Germany, for example, where some science is taught from an integrated perspective up to grades 6, 8, or 10, although only those teachers who studied beyond the primary educational level are educated as chemistry, biology, or physics teachers. Although every teacher in Germany studies two subjects to be taught in school, they are not required to be two science subjects. The teachers can also study (and teach) chemistry combined with maths or any subjects from the social sciences or humanities. When this happens, chemistry teachers face the challenge of teaching biology, earth sciences, and/or physics content without having either studied the subject matter or taken educational courses specific to these domains.

An overview of the great variety of educational systems in Europe, as an example, can be found in a report on the EU's Eurydice project EU (European Commission/EACEA/Eurydice, 2015). For countries outside the EU, Risch (2010) provides a useful overview of similarities and differences in how chemistry is taught and how teachers' pre-service education is organized in 25 countries around the world.

1.2 Approaches to Pre-service Education of Chemistry Teachers

Similar to the between-country, or even within-country differences in educational systems in which chemistry is taught, differences can be found in the pre-service education of chemistry teachers. Pre-service chemistry teacher programs range in length from a 3-year BSc degree in chemistry as the formal qualification to become a chemistry teacher in middle and high schools, to a 7-year integrated chemistry teacher education program with different graduation steps such as, for example, in Germany or Austria. Differences in chemistry teacher education can also be found in the paths to graduation. In general, there are two major models for teacher education, which can be thought of as consecutive and integrated (or concurrent) (Caena, 2014).

In chemistry education, the consecutive approach starts preparing teachers with almost exclusively content-focused chemistry studies at the undergraduate level, leading to a BSc degree in chemistry. The content of the first stage in this model is chemistry and related knowledge, such as physics or maths. In some countries, this qualification is all that is needed to work as a chemistry teacher in middle school, or even at the upper secondary school level. In this case, teachers have to develop their general educational and domain-specific educational skills on the job.

Professional development programs are sometimes offered during a teacher's first years of work. In some countries, these courses are compulsory, in others they are not. More advanced consecutive programs ask prospective teachers who have obtained their BSc degree to enroll in either a teaching certificate program (often 1 or 2 years) or a MEd program (mostly 2 years) before the student teachers become recognized as fully qualified middle- or high-school chemistry teachers (Figure 1.2). These post BSc programs might – but do not always – include school internships and practical teaching exercises. Requirements for completing pre-service teacher education sometimes depend on the type of school or grade levels that the individual will be teaching.

Integrated (or concurrent) approaches to pre-service education start the professional development of the prospective chemistry teachers at the beginning of, or quite early in their undergraduate studies, with a focus on preparing students to become chemistry teachers. Students usually choose to become teachers in their first (or possibly second) year of college/ university, and these students then enroll in courses on both the content of chemistry, with related physics and maths, and general and domain-specific education. School internships and practical teaching experiences are usually integrated into these programs, starting from the undergraduate level. These programs can last from 3 to 4 or 5 years. In Germany, for example, teacher education generally starts with a 3-year program that leads to a BSc. All of the

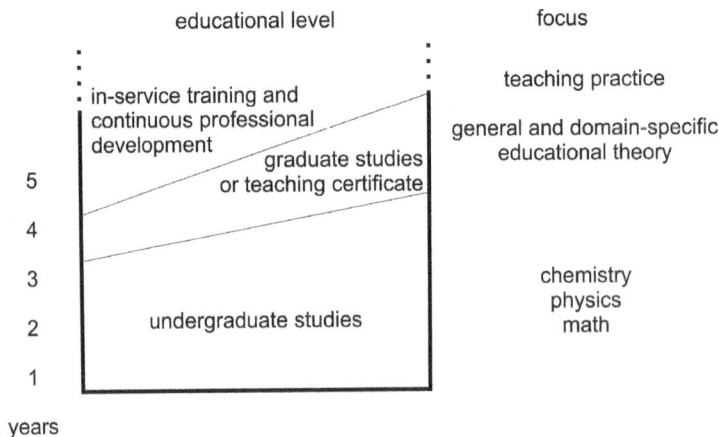

Figure 1.2 Consecutive models of chemistry teacher education.

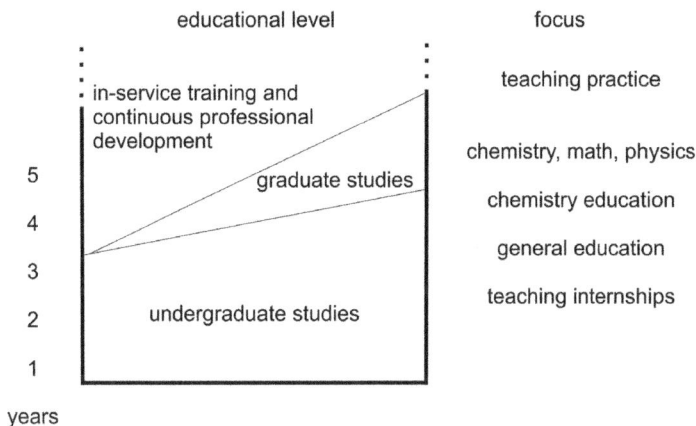

Figure 1.3   Integrative models of chemistry teacher education.

students then spend 2 years in a MEd program followed by 18–24 months of compulsory post-MEd training in schools. The students must pass exams at each of the three steps during these 7 years. All three degrees are then required to become recognized as a fully qualified middle- or high-school teacher (Figure 1.3).

The consecutive and integrated approaches both have advantages and disadvantages. The consecutive programs ensure an in-depth education in the content matter of all of the basic fields of chemistry. The students are "socialized" as chemists to become experts and ambassadors for their subject. The consecutive models allow the student to postpone the decision of whether to work as a chemist or teacher. The consecutive models, however, often limit the amount of instructional time devoted to the study of general and domain-specific education to 1 year of educational studies or courses on the job, or even less. The consecutive models do not allow for integrated learning of the basic chemistry content with an understanding of how it needs to be transformed and conducted in an educational setting (see Chapter 2 of this book). The integrated programs enable students to build connections between the chemistry content knowledge and their understanding of both pedagogy in general, and chemistry-domain-specific pedagogical knowledge in particular. The integrated programs can allow the student teachers to reflect more specifically on the relevant teaching content of the school chemistry curriculum, but the programs need to avoid lowering the level of education in the fundamental principles and theories of chemistry in their academic chemistry studies. The parallel learning of educational theory and chemistry content also provides the prospective chemistry teacher with an opportunity to reflect on their own learning processes within the context of the learning theories they encounter while they themselves are learning new chemistry content. However, because integrated approaches only qualify the students to work in educational settings, a later move into scientific research and engineering professions might be difficult.

Differences in pre-service programs obviously have effects on the requirements and contents of continuous professional development.

Many countries have developed ways to help people who have been practicing chemists become certified as qualified chemistry teachers. One example is a special program at the Weizmann Institute of Science in Israel. To educate teachers with advanced degrees in maths and science along with experience in scientific research, a teacher training program was developed to teach science subjects in grades 7–12, including chemistry. The program is designated for students and graduates who have at least a MSc degree in the relevant fields. The teacher training builds on the individual's understanding of the content of chemistry and focuses on developing this understanding into teacher knowledge; it also provides the skills to promote school learners' deep cognitive understanding by acquiring various teaching methods, including research-based learning, problem solving, projects, discussions, peer learning and operating technology-enriched learning environments. The duration of the program is 2 years. The program consists of six courses: (i) introduction to science education, (ii) learning environments, (iii) assessment, (iv) educational psychology, (v) history and philosophy of science, (vi) cognition. After taking these courses, these individuals attend a full year course in didactics in the chosen discipline, such as chemistry.

In the USA, institutions such as Purdue University that have a long history of graduating teachers who specialize in science, technology, engineering, or maths (STEM) courses at the high-school level have developed programs such as the Transition to Teaching (TtT) program that is open to individuals who have at least 3 years as a practicing chemist, engineer, mathematician, *etc.* These individuals take at least six courses that are graduate-level versions of the courses that undergraduate pre-service teachers are required to take. One of the differences between the TtT program at Purdue and the program at the Weizmann Institute is the ability to tailor the courses to individuals in the program to meet specific needs.

In addition to individuals attending in-service teacher education programs to teach at the high-school level, research-intensive chemistry departments across the USA are turning out PhD graduates who take faculty positions at colleges and universities at every level, from local community colleges to research-intensive universities. Traditionally, graduates of these programs have state-of-the-art knowledge of the content of chemistry within the specific domain in which they graduate, such as inorganic or organic chemistry. They have demonstrated the ability to do chemistry, usually with no background whatsoever in either general pedagogical or domain-specific pedagogical knowledge. A little more than 35 years ago, a program was created to produce PhD (and MSc) graduates who had the advanced content knowledge expected of chemistry faculty, a solid background in education courses, and the ability to research the problems of teaching and learning chemistry (Bodner and Herron, 1984). Twenty-five years later, the program's accomplishments were summarized (Bodner and Towns, 2010). At the end of the 2017–2018 academic year, this program's 100th PhD student will graduate. In most other countries, however, college and university teachers in chemistry are qualified only by being educated as fully trained research chemists. Although sometimes a PhD in chemistry is even a prerequisite, formal educational training is generally not required.

As a case in point, an overview of the diversity of teacher education in Europe alone was provided by Caena (2014). A recently published book by Maciejowska and Byers (2015) discusses selected aspects of good practices in chemistry teachers' pre-service education.

1.3   Consequences for Continuous Professional Development

Initial teacher education provides the prospective chemistry teacher with basic knowledge in chemistry and (hopefully) chemistry education. However, such programs only contribute to a limited extent to the knowledge base of teachers (Van Driel *et al.*, 1998). The notion of teacher knowledge first came to prominence in chemistry education a quarter of a century ago, and there has been a plethora of literature on what teachers know and do to carry out their work (Mulholland and Wallace, 2005). By acknowledging teachers' central role in teaching, the movement to enhance teachers' knowledge places the practicing teacher at the heart of attempts to reform classrooms and improve student achievement. Although there is much agreement about the importance of teachers' knowledge, however, there have also been numerous discussions, debates, and concerns regarding how teachers' knowledge is constructed, organized, and effectively used (*e.g.*, Fenstermacher, 1994; Munby *et al.*, 2001; Kennedy, 2002; Kind, 2009).

Chemistry has an ever-changing knowledge base and its aligned pedagogies and instructional techniques develop over time. Many teachers in school systems worldwide completed their training many years (in the order of decades) ago. As a result, their science knowledge and knowledge of important recent developments regarding science teaching (pedagogical knowledge and knowledge of new curricula and learning environments) have become outdated. This consequently inhibits their ability to implement and operate modern teaching approaches that require contemporary scientific and pedagogical knowledge to teach at an appropriate level and with suitable methodology (Van Driel *et al.*, 1998). That is why, as is true for every teaching profession, chemistry teachers need continuing professional development to update both their chemistry content knowledge and the aligned domain-specific pedagogical knowledge (see Chapter 2 of this book). Moreover, even though teachers attend long-term professional development programs, as recommended by the National Research Council (1996) and by science educators (*e.g.*, Loucks-Horsley and Matsumoto, 1999), the results are sometimes less than would be expected if these programs over-emphasized (because these programs over-emphasize?) teachers' pedagogical knowledge, rather than their content knowledge (Taitelbaum *et al.*, 2008) or *vice versa*.

Maciejowska *et al.* (2015, p. 250), with reference to OECD (1998), suggest the following objectives for continuous professional development:

- to update individuals' knowledge of a subject in light of recent advances in the field
- to update individuals' skills, attitudes and approaches in light of the development of new teaching techniques and objectives, new circumstances and new educational research
- to enable individuals to apply changes made to curricula or other aspects of teaching practice
- to enable schools to develop and apply new strategies concerning the curriculum and other aspects of teaching practice
- to exchange information and expertise among teachers and others, *e.g.*, academics, industrialists
- to help weaker teachers become more effective.

Continuous professional development is essential for school chemistry teaching to become meaningful, more inquiry-based, educationally effective, and better aligned with the chemistry of the 21st century and its related (for example) socio-scientific issues (see Chapter 6 of this book). However, it is very important to take the pre-service qualification and the corresponding knowledge into account when planning professional development programs (Haney *et al.*, 2002). This aspect is also part of a list of quality criteria for professional development stated by Richardson (2003). In general, professional development should:

- be state-wide
- be long term with follow-up
- encourage collegiality and foster agreement among participants on goals and visions
- acknowledge participants' existing beliefs and practices
- have a supportive administration and have access to adequate funds for materials
- involve speakers from outside the school environment.

Although not always possible, the most promising strategies for sustainable change in teaching chemistry require effective and long-term strategies of professional development, including a connection to teachers' prior knowledge and practical experience (see Chapters 2–4 of this book). Providers of professional development need to take into account the knowledge and skills that the teachers bring with them when they attend professional development courses. The best professional development occurs in an environment characterized by multiple exchanges between practitioners, with external experts in the specific educational domain, and new developments in chemistry and its related applications. Chemistry teachers should become familiar with new ideas in chemistry and also understand the implications for themselves as teachers and for their students in the classroom. Only this will allow them to adopt and adapt them for use in their own classrooms.

However, the issue of professional development of chemistry teachers is not dealt with enough in the literature or by policy-makers in chemistry education, although several recent books specifically dealing with the teaching of chemistry now contain relevant chapters (*e.g.*, Mamlok-Naaman *et al.*, 2013; Hugerat *et al.*, 2015; Maciejowska *et al.*, 2015; Van Driel and De Jong, 2015). Because we believe that this is a crucial issue, this book elaborates upon it by describing current programs and studies conducted by the authors. The book also discusses research that has highlighted important features characterizing effective professional development programs, as summarized by Loucks-Horsley *et al.* (1998):

- Engaging teachers in collaborative long-term inquiries into effective teaching practices and student learning.
- Introducing these inquiries into problem-based contexts that consider the content as central and integrate them with pedagogical issues.
- Enabling teachers to approach teaching–learning issues embedded in real classroom contexts through reflections and discussions of each other's teaching and/or examination of students' work.
- Focusing on the specific content or curriculum that teachers will be implementing so that teachers will be given adequate time to determine what and how they need to adapt their current teaching practices.

Mamlok-Naaman *et al.* (2013) argued that characteristics that identify "best practice for teaching" can be recognized by every person who has been in the school context, be it as a pupil, a parent or a teacher. General factors such as "knowing the subject and liking children" might be recognized by everyone (Kind, 2009). It is important to recognize, however, that different exemplary teachers can teach in different ways and still be a source of ideas for "best practice". Some of these differences are related to the subject matter that a particular teacher teaches. There is something about being an exemplary chemistry teacher that is fundamentally different from the "best practices" for being a literature teacher, for example. Therefore, we may consider teaching as a professional activity that is based on a group of actions that are intentionally anticipated by a teacher to promote conceptual, procedural, and attitudinal learning in school. It is also clear, however, that every chemistry teacher has the potential to become a better teacher, no matter how long they have been teaching. This happens by getting more experience, reflecting on their professional work, and continuing the process of being a life-long learner of both the results of chemistry research and newly developed practices in the domain of chemistry education.

Teachers have repeatedly been described as the key to any sustainable reform or innovation in educational practices in general (Hattie, 2008), and in chemistry teaching and learning in particular (Mamlok-Naaman *et al.*, 2013). In this book, we will focus on the critical role of professional development in attaining the goal of high-quality, effective education in chemistry, regardless of the level at which the course is taught. The crucial role of teachers' professional development is highlighted in the literature on the search for relevant chemistry education (*e.g.*, Hugerat *et al.*, 2015). This is why the framework of reforms in chemistry education advocates the existence of extensive, dynamic, and long-term professional development of chemistry teachers. By attending professional development programs, chemistry teachers become acquainted with new developments in chemistry and updated curricular materials, as well as innovative teaching strategies. They also undergo proper professional preparation to implement new curricular materials, and continue to obtain the necessary guidance and support while implementing new curricula that will include new content and its related pedagogies.

1.4   About This Book

There are various models for teachers' professional development, and both teachers and developers of these programs should choose those that match the teachers' needs best at that particular stage in their career and are in alignment with the educational environment in which they operate. This book provides an overview and select cases.

The authors of this book have many decades of experience in the professional development of chemistry teachers in Israel, Germany and the USA. From this rich experience, selected aspects of, and personal contributions to the professional development of teachers are discussed, reflected upon, and illustrated using cases from pre- and in-service chemistry teacher education.

In Chapters 2–5, we discuss the knowledge base of chemistry teachers and a model for understanding teachers' professional development. The chapters examine different approaches and models for the professional development of chemistry teachers in a variety of contexts, including curriculum-implementation programs, teachers as curriculum developers, evidence-based professional development, and the improvement of teaching practice by action research. In Chapters 6–8, contemporary issues of chemistry teachers' professional development are taken into account, including teaching for society, sustainability, and relevant chemistry education, effective teaching in the chemistry laboratory, and the challenges of ever-changing developments in information and communication technologies. Chapter 9 provides guidance on how to prepare chemistry teachers to become educational leaders, and Chapter 10 summarizes the general ideas in this book and outlines further directions.

All of the chapters elaborate on theories and content based on example programs and workshops from the authors' experiences. The discussion is illustrated by examples of the different contents and methods from recent practices. We hope the experiences described in this book will inspire teacher educators in forming initiatives to promote chemistry teacher professional development in their various countries within the context of improving the learning of chemistry by their students.

References

Bodner G. M. and Herron J. D., (1984), Completing the program with a Division of Chemical Education, *J. Coll. Sci. Teach.*, **14**(3), 179–180.

Bodner G. M. and Towns M. H., (2010), The Division of Chemical Education revisited, 25 years later, *J. Coll. Sci. Teach.*, **39**(6), 38–43.

Caena F., (2014), *Initial Teacher Education in Europe: An Overview of Policy Issues*, Brussels: EU Commission, retrieved from http://ec.europa.eu/dgs/education_culture/repository/education/policy/strategic-framework/expert-groups/documents/initial-teacher-education_en.pdf.

European Commission/EACEA/Eurydice, (2015), *The Structure of the European Education Systems 2015/16*, Luxembourg: Publications Office of the European Union.

Fenstermacher G. D., (1994), The knower and known: the nature of knowledge in research on teaching, *Rev. Res. Educ.*, **20**, 3–56.

Haney J. J., Lumpe A. T., Czerniak C. M. and Egan V., (2002), From beliefs to actions: the beliefs and actions of teachers implementing change, *J. Sci. Teach. Educ.*, **13**, 171–187.

Hattie J., (2008), *Visible Learning*, London: Routledge.

Hugerat M., Mamlok-Naaman R., Eilks I. and Hofstein A., (2015), Professional development of chemistry teachers to teach relevant oriented chemistry, in Eilks I. and Hofstein A. (ed.), *Relevant Chemistry Education – From Theory to Practice*, Rotterdam: Sense, pp. 369–386.

Kennedy M. M., (2002), Knowledge and teaching, *Teach. Teach. Theory Pract.*, **8**, 355–370.

Kind V., (2009), Pedagogical content knowledge in science education: perspectives and potential for progress, *Stud. Sci. Educ.*, **45**, 169–204.

Loucks-Horsley S., Hewson P. W., Love N. and Stiles K. E., (1998), *Designing Professional Development for Teachers of Science and Mathematics*, Thousand Oaks: Corwin Press.

Loucks-Horsley S. and Matsumoto C., (1999), Research on professional development for teachers of mathematics and science: the state of the scene, *Sch. Sci. Math.*, **99**, 258–271.

Maciejowska I. and Byers B., (2015), *A Guidebook of Good Practice for the Pre-service Training of Chemistry Teachers*, Krakow: Jagiellonian University.

Maciejowska I., Ctrnatova H. and Bernard P., (2015), Continuing professional development, in Maciejowska I. and Byers B. (ed.), *A Guidebook of Good Practice for the Pre-service Training of Chemistry Teachers*, Krakow: Jagiellonian University, pp. 249–263.

Mamlok-Naaman R., Rauch F., Markic S. and Fernandez C., (2013), How to keep myself being a professional chemistry teacher? in Eilks I. and Hofstein A. (ed.), *Teaching Chemistry – A Studybook*, Rotterdam: Sense, pp. 1–36.

Mulholland J. and Wallace J., (2005), Growing the tree of teacher knowledge: ten years of learning to teach elementary science, *J. Res. Sci. Teach.*, **42**, 767–790.

Munby H., Russell T. and Martin A. K., (2001), Teachers' knowledge and how it develops, in Richardson V. (ed.), *Handbook of Research on Teaching*, New York: Macmillan, pp. 433–436.

National Research Council (1996), *National Science Education Standards*, Washington: National Academy Press.

OECD, (1998), *Staying Ahead: In-Service Training and Teacher Professional Development*, Paris: OECD.

Richardson V., (2003), The dilemmas of professional development, *Phi Delta Kappa*, **84**(5), 401–406.

Risch B., (2010), *Teaching Chemistry Around the World*, Münster: Waxmann.

Taitelbaum D., Mamlok-Naaman R., Carmeli M. and Hofstein A., (2008), Evidence-based continuous professional development (CPD) in the inquiry chemistry laboratory (ICL), *Int. J. Sci. Educ.*, **30**, 593–617.

Van Driel J. H. and De Jong O., (2015), Empowering chemistry teachers' learning: practices and new challenges (pp. 99–122), in Garcia-Martinez J. and Serrano Torregosa E. (ed.), *Chemistry Education*, Weinheim: Wiley-VCH, pp. 693–734.

Van Driel J. H., Verloop N. and de Vos W., (1998), Developing science teachers' pedagogical content knowledge, *J. Res. Sci. Teach.*, **35**, 673–695.

CHAPTER 2

Understanding the Cognitive and Affective Aspects of Chemistry Teachers' Learning and Professional Development

Teaching chemistry is a professional activity based on the knowledge, skills, attitudes, and beliefs of the teacher. Developing chemistry teachers' knowledge and skills, as well as changing their attitudes and beliefs, requires an understanding of the knowledge that chemistry teachers already possess and how it develops. This chapter provides a basis for understanding chemistry teachers' knowledge base and its professional development. It also discusses an example of course-based professional development of chemistry teachers that focused on promoting their knowledge and skills in using the history of chemistry in chemistry teaching.

2.1 Teachers' Professional Knowledge Base

Teachers' professional knowledge base can be defined in terms of the knowledge that they use in their teaching. This knowledge can be divided into content knowledge (CK), general pedagogical knowledge (PK), and domain-specific pedagogical content knowledge (PCK) (Shulman, 1986). Shulman (1987) describes PCK as "that special amalgam of content and pedagogy that is uniquely the province of teachers, their own special form of professional understanding" (p. 15). Shulman (1987) goes on to suggest:

> The key to distinguishing the knowledge base of teaching lies at the intersection of content and pedagogy, in the capacity of a teacher to

Advances in Chemistry Education Series No. 1
Professional Development of Chemistry Teachers: Theory and Practice
By Rachel Mamlok-Naaman, Ingo Eilks, George Bodner and Avi Hofstein
© Rachel Mamlok-Naaman, Ingo Eilks, George Bodner and Avi Hofstein 2022
Published by the Royal Society of Chemistry, www.rsc.org

transform the content knowledge he or she possesses into forms that are pedagogically powerful and yet adaptive to the variations in ability and background presented by the students. (1987, p. 15)

Shulman's idea of PCK was later expanded upon by Van Driel *et al.* (1998) as a specific form of craft knowledge, and by Loughran *et al.* (2001) as: "the knowledge that a teacher uses to provide teaching situations that help learners make sense of particular science content."

According to Shulman (1986, 1987), teachers' competence in the areas of general educational knowledge (PK) and subject matter knowledge (CK) does not necessarily lead to the ability to structure student-friendly content and pedagogy to teach science. Shulman (1986, 1987) suggested taking a deeper look into the areas connected to PCK if we want to better understand and develop a teacher's repertoire of professional actions. He claimed that teachers need strong PCK to be the best possible teachers they can be in their specific subject. Loughran *et al.* (2001) described this need as that which teachers must know to best teach content to students. PCK became increasingly accepted as a heuristic device for understanding this specific domain of teachers' knowledge and beliefs relating to their classroom practices in a certain subject (Gess-Newsome and Lederman, 1999).

By definition, PCK is subject- and domain-specific (Shulman, 1986, 1987). Geddis (1993) emphasized that:

The outstanding teacher is not simply a 'teacher', but rather a 'history teacher', a 'chemistry teacher', or an 'English teacher'. While in some sense there are generic teaching skills, many of the pedagogical skills of the outstanding teacher are content-specific. Beginning teachers need to learn not just 'how to teach', but rather 'how to teach electricity', 'how to teach world history', or 'how to teach fractions'. (p. 675)

Bucat (2004) expanded on this by adding: "... how to teach 'stoichi-ometry', 'how to teach chemical equilibrium', 'how to teach stereochemistry'."

PCK can also reflect the differences between teachers' knowledge and the knowledge of content experts, such as practicing scientists or engineers. Cochran *et al.* (1991) described this difference as:

Teachers differ from biologists, historians, writers, or educational researchers, not necessarily in the quality or quantity of their subject matter knowledge, but in how that knowledge is organized and used. For example, experienced science teachers' knowledge of science is structured from a teaching perspective and is used as a basis for helping students to understand specific concepts. A scientist's knowledge, on the other hand, is structured from a research perspective and is used as a basis for the construction of new knowledge in the field. (p. 5)

The concept of PCK suggests that there cannot be one common strategy to be operated on in different contexts, such as in the teaching of language, history, or science. Teachers need to select, adopt, and adapt teaching approaches and methods for each specific situation they encounter with regard to the subject they teach and the learners enrolled in their classes. This process requires domain-specific knowledge on: (i) subject matter and the curriculum (Van Driel *et al.*, 1998), (ii) alternative students' understanding (Lederman *et al.*, 1994), and/or (iii) any available experiments, models, and teaching concepts developed by other teachers (De Jong *et al.*, 2005).

Using a more generalizable approach, Magnusson *et al.* (1999) described the essential domains of PCK as orientation with respect to teaching, knowledge of the curriculum, knowledge of the testing of knowledge, knowledge about learners, and knowledge about strategies of passing on knowledge. For chemistry education, this means knowing about:

- orientations and emphases toward chemistry teaching
- knowledge and beliefs about the chemistry curriculum
- knowledge and beliefs about students' understanding of chemistry topics
- knowledge and beliefs about assessment in chemistry
- knowledge and beliefs about instructional strategies and pedagogies for teaching chemistry.

PCK is developed based on different sources. Grossman (1990) identified four sources from which PCK is generated, and Appleton and Kindt (1999) added a fifth one:

- observation of classes, both as a student and as a student teacher, often leading to tacit and conservative PCK
- disciplinary education in pre-service content courses, which may lead to personal preferences for specific purposes or topics
- specific courses during teacher education, the impact of which is normally unknown
- classroom teaching experience
- recommendations from trusted colleagues.

This means that any teacher's PCK is a very personal set of knowledge and beliefs. PCK is developed and constantly refined on the basis of external input and individual experiences. It is influenced by beliefs, including epistemological beliefs, general educational beliefs, content-related beliefs, beliefs about curriculum orientations with their corresponding emphases, and much more. As a result, PCK is intangible and rarely made explicit by teachers. PCK comes to the fore when teachers act or start to explain what they are doing in the classroom. Consequently, teachers' active presentation of their teaching practices should be an essential component in any professional development program.

2.2   Understanding the Development of Teachers' Knowledge Base

Professional development can be defined as the process of change in a teacher's knowledge, mental models, attitudes, personal beliefs, and/or perceptions concerning teaching and learning (Richardson and Placier, 2001). However, pre-service education can only provide the prospective teachers with a limited amount of (mainly theoretical) knowledge. Over the course of a teacher's career, Berliner (1988) proposed an ideal of five stages of teacher professional development, as follows:

- Novice (classroom teaching is rational and relatively inflexible).
- Advanced beginner (teachers develop strategic knowledge and classroom experiences, and the contexts of problems begin to guide their behavior).
- Competent (teachers make conscious choices about actions, know the nature of timing and what is and is not important).
- Proficient (intuition and know-how begin to guide performance and a holistic recognition among contexts is acquired; teachers can predict events).
- Expert (intuitive grasp of situations, teaching performance is fluid as the teachers no longer consciously choose the focuses of attention).

Berliner (1992) suggested that teachers become better professionals by experience and learning over time to teach, *e.g.*, "school chemistry". They develop and adopt a bigger repertoire of instructional strategies related to a specific topic and know about the difficulties students might have related to that topic. The extended repertoire about how to deal with these difficulties in the classroom and transform the difficulties in learning opportunities for those particular students helps them in practice. This knowledge is developed over time and the result of pre-service training, in-service experiences, and information from external sources. In the life of a teacher, knowledge is continuously changing and being reorganized. This change contributes to the different steps in a teacher's professional development.

Sprinthall *et al.* (1996) outlined three interpretations when distinguishing between different models explaining professional development: craft, expert, and interactive. The craft interpretation is based on the view that teachers develop as a result of gaining experience. Knowledge emerges from classroom experiences, although it is exactly how teachers arrive at their new interpretations and why some teachers reproduce the same experience countless times without any learning taking place. The expert interpretation views teachers as relatively passive consumers, who are taught and trained by education experts. The interactive interpretation of development models explains professional growth in terms of increasing expertise and growing expertise. It recognizes the fact that professional growth is an interaction between both training and experience.

One of the most recognized models of professional development in recent years has been exactly such an interactive model, the Interconnected Model of Teacher Professional Growth (IMTPG) by Clarke and Hollingsworth (2002). Application of the IMTPG as an analytical tool to reflect teachers' professional development is described for its specific use in science education by, for example, Van Driel and Justi (2006) and Mamlok-Naaman and Eilks (2012), as will be seen in Chapter 5 of this book.

The IMTPG reminds us that the teacher becomes the learner in professional development processes. Thus, we must consider the basic theories of learning and the role of influential factors in a successful and sustainable learning process when attempting to affect teachers' learning. Teacher training following the IMTPG consists of a process based on self-reflection and action that is determined by four domains:

- the personal domain (beliefs, attitudes, and earlier experiences)
- the practical domain (the teacher's authentic teaching practices)
- the external domain (topic requirements, media, and curriculum aspects)
- the domain of consequences (goals and effects).

The IMTPG suggests that a change in any one of the four domains is translated into the other domains through the mediating processes of enactment and reflection, and that teachers' (ongoing) professional development is not based on a persistent, structural process of top-down information delivery. Effective professional development needs to be connected to explicating the teachers' beliefs, attitudes, and experiences, to facilitate their making meaning of their own experiences, and finally to help them learn about the effects.

Each of the IMTPG domains has its own set of subdomains, which can be differentiated into factors such as CK, PK, and PCK, as noted by Shulman (1986, 1987). Even though it seems clear that CK, PK, and PCK cannot be sharply detached from one other, differentiation may help us give more structure and meaning to the different domains in the IMTPG. These domains will, of course, interact when it is time for the teachers to actually practice their trade (Tobin *et al.*, 1994; Magnusson *et al.*, 1999).

Considering the four domains of the IMTPG (Clarke and Hollingsworth, 2002), one easily recognizes that a teacher's ongoing professional development should not be solely based on a persistent, structural process of information and training delivery. Researchers know that teachers' learning is only partially based on specific courses during their teacher education. Learning is influenced at least as much by direct observation of classes – both as a student and as a student teacher – and by both classroom teaching experience (Grossman, 1990) and recommendations from trusted colleagues (Appleton and Kindt, 1999). Therefore many, if not most, central aspects of a teacher's professional growth are developed unconsciously. These touch upon the knowledge domain, but they also draw on beliefs and attitudes toward educational aspects in general, including domain-specific aspects bound to the personal PCK of the teachers. Such aspects can be assumed to exist for all four IMTPG domains, but especially for the personal and practical ones. An example of how to use the IMTPG to reflect teachers' professional development is presented in Chapter 5 of this book.

In the final analysis, teachers require innovative approaches that support their own personal learning in the domains of CK, PK, and PCK. These will strengthen their ability to recognize and utilize their personal and practical beliefs, make such beliefs explicit, and aid them in the process of self-reflection. Huberman (1993) stated that persistent, long-term interaction with people from outside the specific school setting is unavoidable in achieving research-based innovation in teaching practices and professional development. This process must be centered on collective reflections stemming from both the teachers themselves and their external partners with respect to current and altered teaching practices (*e.g.*, Haney *et al.*, 1996).

2.3   Teachers' Need for Lifelong and Intense Professional Learning

As already noted, Huberman (1993) argued for long-term interaction of teachers with people from outside their corresponding schools to achieve research-based innovation and professional development. This cooperation should include joint reflections of both the teachers and external partners with respect to current and altered teaching practices (*e.g.*, Haney *et al.*, 1996), as well as the process of professional development (*e.g.*, Eilks and Markic, 2011). Such joint reflections may help promote teacher learning by incorporating a social-constructivist and situated learning perspective.

Abell (2007) stressed that the situated learning perspective could be well applied to teacher learning. This means that teachers' learning is directly related to their professional environment and their professional communities. The role of every teacher therefore involves being simultaneously a learner and a partner in the development of other teachers' professional practices. Thus, every teacher also has the responsibility of helping other teachers and novice colleagues reach the level of expertise that he or she already has. At the same time, they improve their own teaching by working as a mentor for a colleague or novice teacher.

However, not every piece of information is available within one's own school environment. Teachers need to continuously familiarize themselves with new approaches to teaching chemistry, for example based on emerging new technologies, new content, or advancing pedagogies, and with ideas and methods for better practice from the whole community of chemistry education. Examples of new technologies include teaching with digital and social media (see Chapter 8 of this book); examples of new content and perspectives are nanochemistry or sustainability and green chemistry (see Chapter 6 of this book). For changes in pedagogy, see, for example, Eilks *et al.* (2013).

Although this is the ideal picture, teachers also have to learn about recent advances in evidence from educational research, and particularly in the case of chemistry teachers, chemistry education research. The teachers need to understand and reflect upon the implications of newly gained evidence for themselves as teachers and for their learners in the classroom before they adopt and adapt research findings and their implications. This could be more efficiently done with colleagues from their own regional school environment and, if available, external experts from chemistry education research as a collaborative learning process (see Chapter 5 of this book). If new evidence or teaching approaches differ greatly from the teachers' previous practice, joint reflections can help reshape teachers' beliefs regarding the practice of chemistry teaching and learning. The reflection should involve careful consideration of core principles and issues from research, as well as their contextualization in the process of developing practice. For such a process to occur, teachers' input from outside their own school or from other subjects within their schools, as well as support from outside the school environment, can be of benefit.

As top-down, short, and occasional professional development activities are of limited potential (Loucks-Horsley *et al.*, 1998), bottom-up, long-term, or structured follow-up approaches are needed (Arce *et al.*, 2014). In general, effective programs for teachers' lifelong learning should respect the following strategies (*e.g.*, Loucks-Horsley *et al.*, 1998; Marx *et al.*, 1998; Taitelbaum *et al.*, 2008):

- Engaging teachers in collaborative long-term inquiries into teaching practice and student learning.
- Situating these inquiries in problem-based contexts that place content as central and integrated with pedagogical issues.
- Enabling teachers to see such issues as being embedded in real classroom contexts through reflections and discussions of each other's teaching and/or examination of students' work.
- Focusing on the specific content or curriculum that teachers will be implementing so that the teachers are given time to work out what and how they need to adapt what they are already doing.

In general, we can say that effective continuous professional development programs for teachers (Lipowsky, 2010):

- are long term in nature and enable a more in-depth discussion of content because short-term professional development initiatives seem to be unable to impact teaching routines
- are characterized by a variety of methodological settings (alternation of input and work phases, experimentation, training, and reflection sequences)
- clearly tie in with the participant's classroom practice (*e.g.*, start out with concrete challenges)

- place a domain-specific educational focus on selected issues and allow teachers to address contents in depth
- encourage critical questioning of teachers' fundamental convictions and therefore form the basis for a lasting change in attitudes and classroom practice (including a systematic reflection of one's own practice and of the underlying assumptions)
- explicitly provide for cooperation among teachers beyond the event (*e.g.*, promoting and using regional networks)
- involve several teachers from a given school and encourage in-school dissemination of the course contents
- offer external support for the in-school implementation of the course contents.

Effective professional development that results in a change in teachers' professional knowledge needs to provide opportunities for teacher reflection and for learning about how new practices can evolve or be shaped from existing classroom practice. This is not a simple task, in that it requires teachers to re-examine what they do and how they might do it differently.

The following case describes an example of a structured approach to teachers' professional development that focused on incorporating the history of chemistry into chemistry education.

2.4 Chemistry Teachers' Professional Development in the Framework of Curriculum Innovation to Incorporate the History of Chemistry into Teaching

2.4.1 The Starting Point

With the goal of teaching *Science for All*, the faculty and staff at the Weizmann Institute of Science developed an initiative for curriculum reform and teacher professional development based on a teaching and learning module named *Science – an Ever-Developing Entity* (Mamlok-Naaman *et al.*, 2005). The module, which was intended for non-science majors in Israeli high schools (10th-graders), was designed to develop an understanding of the nature of science among all students by using historical examples. In the lesson plan, science was presented as a continuously developing enterprise of the human mind illustrated by the historical development of our contemporary understanding of the structure of matter.

The reason for this initiative was that even today, many high-school students finish their formal education believing that science is an enterprise in which some smart and studious individuals (scientists) discover the true facts of reality. Many students, even those who intend to become scientists, are unaware of the true nature of science (Duschl and Grandy, 2013). Students, both in high school and at the college level, often hold purely positivistic views regarding the nature of physical reality and scientific inquiry (Erduran and Dagher, 2014). While students may understand science as a systematic gathering of facts and laws, they may not be aware of the roles of science and scientists in building models and theories as tools to explain nature (Jungwirth, 1987).

Rarely is science perceived by students as the process by which people construct a satisfying grasp of natural phenomena, a process that involves endless testing and refining of the plausibility of their solutions. Nevertheless, a goal of science education is that all citizens understand the process of science. As the scientific community debates these solutions, models, and theories, new knowledge is developed that eventually becomes the best contemporary understanding of nature. The goal of the professional development activity described in this section was to help teachers understand how to get their students to realize that knowledge created by scientists is being continually refined and rethought, through the suggestion and performance of new experiments and measurements.

Arons (1984) claimed that many science teachers downplay the scientific process and, as a result, miss opportunities to teach critical and investigative thinking. Furthermore, they fail to emphasize that science develops culturally according to the "spirit" of each historical period, and is related to technological, political, sociological, and cultural developments (Lederman *et al.*, 2014). If science is truly to be a subject learned by all, students must perceive it as a continuously developing product of the human mind and appreciate it as a vital program for non-science majors (Project 2061, 1989).

2.4.2 Cognitive Aspects of the Historical Approach

McCloskey (1983) claimed that the stages in which children develop their scientific thinking parallel the stages of the historical development of science from the beginning of time; in other words, "ontogeny recapitulates phylogeny". In this view, a child's first concepts of science are similar to the concepts of scientists in ancient times, when human qualities were conferred upon inanimate objects and natural processes were described in terms of human emotions. Thus, it is not surprising that young students build their conceptual world according to their own knowledge and feelings. Like their ancestors before them, these beliefs are based on their feelings, senses, and understanding of the world around them. For example, children cannot conceive of gases having any weight, because they are invisible (Furio Mas *et al.*, 1987). Similarly, children assume that if materials are a collection of particles, then the properties of each particle, atom, or molecule are the same as those of the substance on the macroscopic level, *e.g.*, individual gold atoms have the same color as a macroscopic sample of gold. The theory of atoms and particles that describes solid matter as mostly empty space in which particles move in a vacuum was not accepted until the 17th century. It contradicted sensual perceptions and the desire for harmony (Matthews, 1994). Nevertheless, few students (or even adults) conceive of "solid" matter as primarily empty space.

Ben-Zvi *et al.* (1986) noted that the difficulties faced by students in adopting the particulate nature of matter are not surprising, because it took mankind almost 2000 years to develop and accept this model of the macroscopic world in which we live. If we assume that the naïve models that students bring to class are part of a normal cognitive evolution, then showing students why and how our scientific models have changed may help them replace their ancient, simplistic models with more modern and complex ones.

Conceptual changes are usually accompanied by cognitive as well as affective stress and difficulties. A historical approach enables us to demonstrate the development of theories and shows that science is an ever-developing enterprise where scientists routinely modify their ideas through logic, experiment, and experience (Ben-Zvi *et al.*, 1986; Abd-El-Khalick and Lederman, 2000). Hall *et al.* (1983) have shown that a historical approach may help students overcome some learning difficulties while leading to a better grasp of the concepts involved.

2.4.3 The Module: Science – An Ever-developing Entity

Because the major cognitive goal of the module was to teach the nature of science (while encouraging science-alienated students to continue their study of science), we identified five principal objectives to guide its development. We wanted students completing this module to understand that:

- What sets us apart from other members of the animal kingdom is our ability to ask questions about ourselves and the world around us.
- Science is a continuous, ongoing process of questioning and searching for answers and is, therefore, an essential part of our cultural heritage.
- The development of our understanding of the structure of matter exemplifies the main features of the scientific approach.
- There is a continuous interplay between advancing science knowledge and technological developments.
- The most essential questions remain basically the same, although the answers have changed throughout history. These changes are influenced by the available technological devices in a given period as well as by the concurrent social norms and pressures.

As the primary goal of the lesson plan was to enhance students' understanding of the nature of science, we designed a module that systematically looks at various science concepts across time. However, having decided on the historical approach, we still faced many problems and we had to ask ourselves:

- What are the dangers of a chronological approach? What if they are inconsistent with models assumed to be correct in the intervening years?

- How should one introduce the main concepts and issues needed to understand the structure of matter? How much background information do students need? How should we provide it without introducing additional misconceptions?

At the same time, our goal was to integrate issues that deal with society, economics or culture, without forgetting the main goal of the lessons – teaching a particular scientific topic – and to determine the extent to which one can introduce the life stories of scientists without making the teaching too narrative.

The module developed around two interwoven issues: the interrelationship between theories and experimental data on the one hand, and the links that exist between science and technology on the other. We wanted to include interesting topics to motivate students while leading to our goals, and two topics seemed to be appropriate: (i) the discovery of electricity and (ii) the possibility of transmuting base metals into gold. Figure 2.1 shows the structure of the module.

The lessons traced changes in the way the question "Can base metals be transmuted into gold?" was approached and answered during the history of science. In doing so, it illustrated historical changes in our understanding and our current conceptions of the structure of matter. These changes across time reflect the interplay between facts and theories, as well as the interaction of culture and prevailing thought. In the Ancient Greek period, matter was conceived of as just four entities or elements – earth, water, air, and fire. Each element differed in terms of the combination of two properties: hot *versus* cold and dry *versus* wet. Air, for example, was hot and wet whereas water was cold and wet. The various elements were mixed and/or separated by two emotions: love and hate. Each element sought its preferred place, which is why fire and air rise and earth and water fall. Each of these elements could be changed into another by performing different operations, such as heating, cooling, mixing, crystallizing, *etc.* The natural conclusion, based on this theory, was that gold could be made from other materials provided that one was clever enough to know the right transforming operations and sequence.

The lessons described how the introduction of quantitative considerations into the chemistry laboratory brought about a revival of the atomic theory originally postulated by the Greeks. Matter, according to Dalton's 18th century views, was made up of small, indestructible particles. Dalton's atoms were the basic units of this matter, hence they could not be interchanged. Because Dalton's theory was generally accepted, scientists (and even alchemists) concluded that one element could not be transformed into another. Thus ended the period of ancient alchemy (Conant and Nash, 1957).

More recent experimental data, such as from natural and artificial nuclear reactions, provide a deeper understanding of the structure of matter. Atoms are no longer considered the ultimate basic particles of matter and hence, elements can be changed into other elements (although only through nuclear reactions). Can we therefore call modern physicists the modern

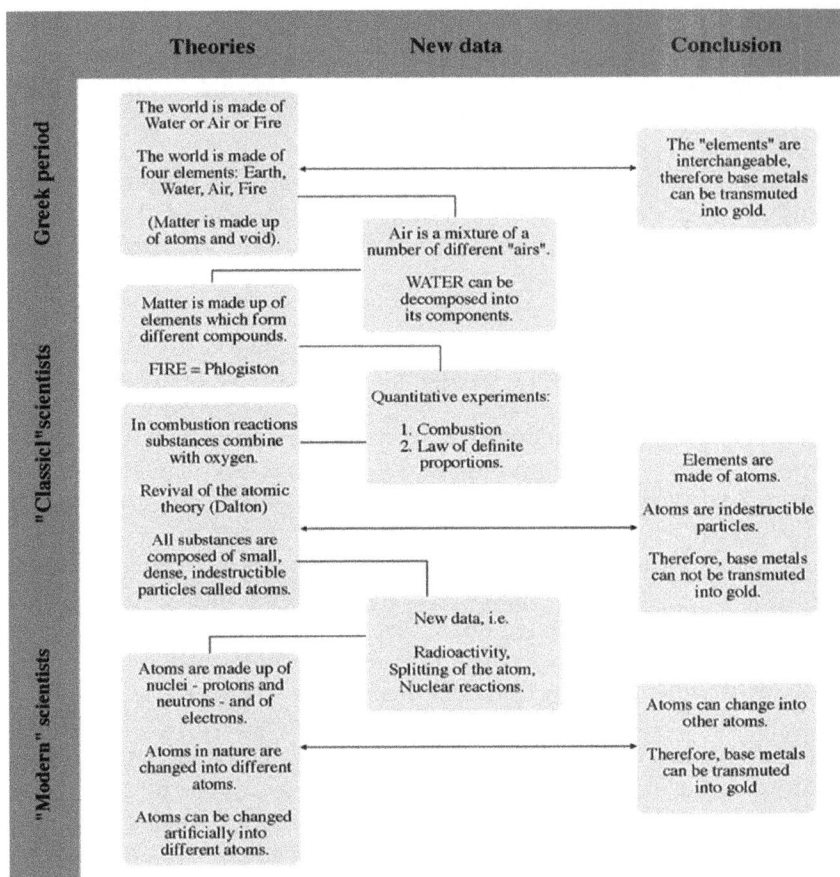

Figure 2.1 The structure of the teaching module "Science – an Ever-Developing Entity".

alchemists? After all, we now know how to produce gold by using this method! However, although possible, economic calculations show that it is far cheaper to exploit gold from natural sources than to obtain it through transmutational methods. Although the modern alchemists may have fulfilled the vision of their ancient counterparts, it required many years of experimentation, knowledge generation, and a revolutionary change in the fields of science and technology. Figure 2.2 presents the relationship between science and technology.

This representation of the process of discovery was chosen to show that:

- Science is not simply a mere collection of facts discovered randomly in the minds of scientists.
- The ancient alchemists, strange and improbable as their ideas may seem to us now, worked within the framework of their contemporary theories, just as scientists do today.

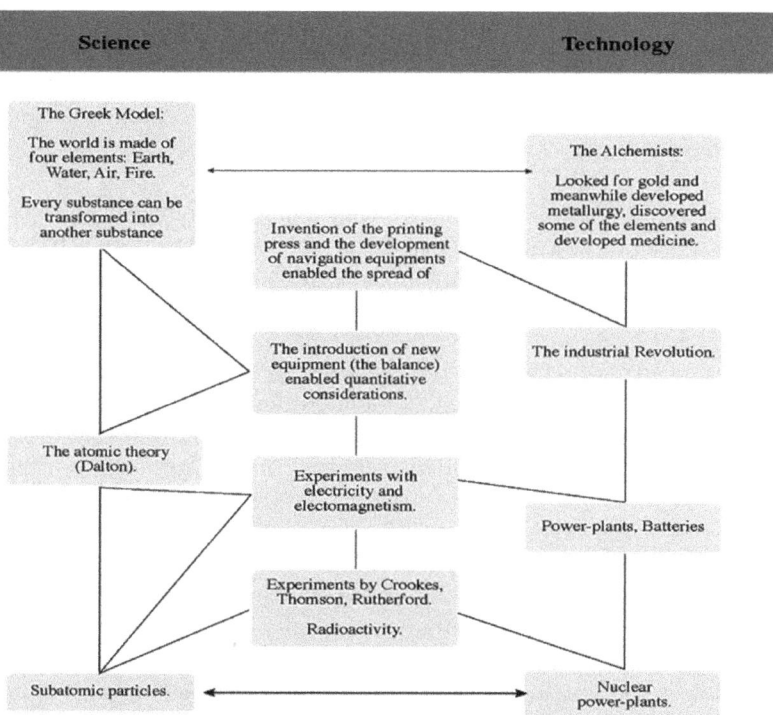

Figure 2.2 The relationship between science and technology.

- Scientists wish to understand how and why things are as they are, in order to predict and then control what happens. The outcomes of their efforts may be used for the benefit of mankind or, conversely, as a tool for its destruction. The decision of how to use science and scientific understanding is in our hands.

2.4.4 Professional Development

The lesson plan, designed for 40 class periods of 45 minutes each, used a historical approach to emphasize links that have always existed between science and technology. In this way, students might understand how technology influenced theory, and vice versa. For example, the introduction of new equipment, such as the balance, enabled quantitative considerations that led to the revival of the atomic theory. On the other hand, theoretical developments of the atomic model enabled the development of batteries and the wide use of electricity from conventional, and later on, nuclear power plants.

However, to teach chemistry through a historical approach, teachers need to undergo suitable preparation. It requires the teaching of aspects of science with which teachers have little familiarity (Brickhouse, 1991; Gallagher, 1991). In addition, teachers themselves may have misconceptions regarding aspects of the nature of science. For example, in terms of the roles of models and modeling, teachers need to understand how such models relate to macroscopic and microscopic features of substances, as well as how the two relate to each other (Gilbert, 1997). Erduran (2003) claimed that lack of effective communication between students and teachers may lead to a mismatch between what is taught and what is learned. Therefore, it is necessary to promote teachers' PCK in issues including CK, and the understanding of students' prior conceptions (Shulman, 1986).

The lesson plan was field tested during one school year and then revised. Implementation and dissemination after revision was accompanied by intensive, comprehensive, and long-term teacher professional development activities and evaluation studies on students and teachers. The teachers who were involved with teaching the module had initial training in biology, chemistry or physics, and usually taught students who were majoring in the sciences. These teachers were not initially trained to teach interdisciplinary topics, or curricula that included historical or philosophical concepts and principles. Teachers attended training activities related to the history and nature of science, as well as to pedagogical techniques that were appropriate for the lesson plan, emphasizing collaborative study and inquiry approaches to teaching and learning. Teachers were continuously supervised by science education staff from the Weizmann Institute of Science during the course.

Before and during their teaching of the lesson plan, teachers attended 50 hours of lectures, discussions, and workshops dealing with the history and philosophy of science, instructional strategies, and science concepts. The value of integrating historical aspects of science in the high-school curriculum was discussed as well (Erduran *et al.*, 2007). During this time, they often worked in small groups and discussed the interdisciplinary nature of the content, the difficulties of coping with such topics, and the required teaching methods. Relying on their rich experience as high-school teachers, the teachers who participated in those workshops also prepared student supplementary materials for each chapter of *Science – an Ever-Developing Entity*. These materials included worksheets, lab experiments, exercises, guided reading of articles, and games.

Instructional strategies used in the lesson plan included case study methods, analysis of source material, the performance of experiments that were similar or identical to those carried out at different periods in history, discussions, and debates. During the course of study, the high-school students who studied the module conducted projects, watched scientific films, and analyzed relevant scientific articles. A large part of their work was done in small collaborative groups.

2.4.5   Assessment of the Outcomes Incorporating Both Qualitative and Quantitative Methods

Qualitative analysis involved the curriculum developers' close observation of lessons, student and teacher interviews, and describing, analyzing and disserting events. Quantitative measures assessed variables in the students' cognitive and affective domains and allowed comparison of experimental and control groups, specifically of the achievements and attitudes before and after the teaching of science programs in the two groups. The experimental group consisted of 10th-grade students who had not chosen to major in science and who studied the module; the control group consisted of a heterogeneous population of 10th-grade students (some of whom were science majors), who studied science according to the established curriculum, *i.e.*, 3 hours of biology, chemistry, and physics per week.

The analysis was conducted in two phases. A pilot study was conducted at the beginning of the 1995 academic year, prior to the development of the first version of the unit. The main goal of the assessment in 1995 was to obtain information about students' interest in science as well as their reasons for opting out of science studies. Results of the study served as a guideline for module development. Intermediate and summative evaluations were conducted during the 1996 and 1997 academic years. The 1996 assessment, the first year in which the module was taught, compared the suitability of the module to the needs of the students and helped us match evaluation tools to our objectives. Following revisions based on our pilot data and recommendations from teachers, a third version of the unit was written.

In the beginning, it was hoped that teaching chemistry from a historical point of view to high-school students who had not chosen to major in science would improve their attitude toward science in general and scientific studies in particular, as well as their understanding of the essence of science, its development, and concepts related to the structure of matter. Further, the lessons were designed to demonstrate the application of general ideas from the history and philosophy of science to specific topics. The structure of matter was chosen as an example of the development of scientific thinking in terms of improved models and evolving theories. Our data indicated that students were successful at learning science concepts and finished the module with positive attitudes toward the study of science.

Based on our current understanding of teachers using this approach, we recommend that: prior to and during the implementation of a new curriculum, teachers be involved in preparing workshops, including development of auxiliary and learning materials, as part of their professional development, and receive continuous support from the curriculum developers. While not fully supported by our data, we also feel that the pedagogy used to teach a new curriculum should examine a variety of instructional methods shown by research to be effective for learning science and developing positive attitudes when working with non-science students (Mamlok-Naaman *et al.*, 2005).

2.5 Summary

- The professional development of chemistry teachers needs to address all three professional knowledge domains of teachers, namely, update the teachers' chemistry CK, familiarize them with current developments in the PK of teaching and learning, and develop the associated, chemistry-specific PCK.

- Chemistry teachers' professional development needs to take all four domains of teachers' professional growth into account, as described in the IMTPG: the personal domain (beliefs, attitudes, and previous experiences), the practical domain (the authentic teaching practices of the teacher), the external domain (topic requirements, media, and curriculum aspects), and the domain of consequences (goals and effects).
- Ideally, the chemistry teacher's development should involve a long-term commitment, inspired or supported by external experts, with phases of active involvement in the planning and structuring of chemistry lessons, such as illustrated by the example presented herein of involving the history of science in chemistry teaching.

References

Abd-El-Khalick F. and Lederman N. G., (2000), The influence of history of science courses on students' views of nature of science, *J. Res. Sci. Teach.*, 37, 1057–1095.

Abell S. K., (2007), Research on science teacher knowledge, in Abell S. K. and Lederman N. G. (ed.), *Handbook of Research on Science Education*, Mahwah: Lawrence Erlbaum, pp. 1105–1140.

Appleton K. and Kindt T., (1999), *How do beginning elementary teachers cope with science? Development of pedagogical content knowledge in science*, Paper presented at the annual meeting of NARST, Boston, USA.

Arce J., Bodner G. M. and Hutchinson K., (2014), A study of the impact of inquiry-based professional development experiences on the beliefs of intermediate science teachers about "best practices" for classroom teaching, *Int. J. Educ. Math. Sci. Technol.*, 2, 85–95.

Arons A. B., (1984), Education through science, *J. Coll. Sci. Teach.*, 13, 210–220.

Ben-Zvi R., Eylon B. and Silberstein J., (1986), Is an atom of copper malleable?, *J. Chem. Educ.*, 63, 64–66.

Berliner D. C., (1988), *The Development of Expertise in Pedagogy*, Washington: AACTE.

Berliner D. C., (1992), Redesigning classroom activities for the future, *Educ. Technol.*, 32, 7–13.

Brickhouse N., (1991), Teachers beliefs about the nature of science and their relationship to classroom practice, *J. Teach. Educ.*, 41(3), 53–62.

Bucat B., (2004), Pedagogical content knowledge as a way forward. Applied research in chemistry education, *Chem. Educ. Res. Pract.*, 5, 215–228.

Clarke D. and Hollingsworth H., (2002), Elaborating a model of teacher professional growth, *Teaching and Teacher Education*, 18, 947–967.

Cochran K. F., King R. A. and DeRuiter J. A., (1991), *Pedagogical Content Knowledge: A Tentative Model for Teacher Preparation*, East Lansing: NCRTL.

Conant J. B. and Nash L. K., (1957), *Harvard Case Histories in Experimental Science*, Cambridge: Harvard University Press.

De Jong O., Van Driel J. H. and Verloop N., (2005), Preservice teachers' pedagogical content knowledge of using particle models in teaching chemistry, *J. Res. Sci. Teach.*, **42**, 947–964.

Duschl R. A. and Grandy R., (2013), Two views about explicitly teaching nature of science, *Sci. Educ.*, **22**, 2109–2139.

Eilks I. and Markic S., (2011), Effects of a long-term Participatory Action Research project on science teachers' professional development, *Eurasia J. Math. Sci. Tech. Educ.*, **7**, 149–160.

Eilks I., Prins G. T. and Lazarowitz R., (2013), How to organize the classroom in a student-active mode, in Eilks I. and Hofstein A. (ed.), *Teaching Chemistry – A Studybook*, Rotterdam: Sense, pp. 183–212.

Erduran S., (2003), Examining the mismatch between pupil and teacher knowledge in acid-base chemistry, *Sch. Sci. Rev.*, **84**(308), 81–87.

Erduran S., Aduriz-Bravo A. and Mamlok-Naaman R., (2007), Developing epistemologically empowered teachers: examining the role of philosophy of chemistry in teacher education, *Sci. Educ.*, **16**, 975–989.

Erduran S. and Dagher Z., (2014), *Reconceptualizing the Nature of Science for Science Education: Scientific Knowledge, Practices and Other Family Categories*, Dordrecht: Springer.

Furio Mas C. J., Perez J. H. and Harris H. H., (1987), Parallels between adolescents' conceptions of gases and the history of chemistry, *J. Chem. Educ.*, **64**, 616–618.

Gallagher J. J., (1991), Prospective and practicing secondary school science teachers knowledge and beliefs about the philosophy of science, *Sci. Educ.*, **75**, 121–133.

Geddis A. N., (1993), Transforming subject-matter knowledge: the role of pedagogical content knowledge in learning to reflect on teaching, *Int. J. Sci. Educ.*, **15**, 673–683.

Gess-Newsome J. and Lederman B., (1999), *Examining PCK: The Construct and Its Implications for Science Education*, Boston: Kluwer.

Gilbert J., (1997), Models in science and science education, in Gilbert J. (ed.), *Exploring Models and Modelling in Science and Technology Education: Contributions from the MISTRE Group*, Reading: The University of Reading, pp. 5–19.

Grossman P. L., (1990), *The Making of a Teacher: Teacher Knowledge and Teacher Education*, New York: Teachers College Press.

Hall D., Lowe L., McKavanagh C., McKenzie S. and Martin H., (1983), *Teaching Science, Technology and Society in Junior High School*, Brisbane: Brisbane College of Advanced Education.

Haney J. J., Czerniak C. M. and Lumpe A. T., (1996), Teacher beliefs and intentions regarding the implementation of science education reform strands, *J. Res. Sci. Teach.*, **33**, 971–993.

Huberman M., (1993), Linking the practitioner and researcher communities for school improvement, *Sch. Eff. Sch. Improv.*, **4**, 1–16.

Jungwirth J., (1987), The intellectual skill of suspending judgement – do pupils possess it?, *Gifted Educ. Int.*, **6**, 71–77.

Lederman G. L., Antink A. and Bartos S., (2014), Nature of science, scientific inquiry, and socio-scientific issues arising from genetics: a pathway to developing a scientifically literate citizenry, *Sci. Educ.*, **23**, 285–302.

Lederman N. G., Gess-Newsome J. and Latz M. S., (1994), The nature and development of pre-service teachers' conceptions of subject matter and pedagogy, *J. Res. Sci. Teach.*, **31**, 129–146.

Lipowsky F., (2010), Lernen im Beruf – Empirische Befunde zur Wirksamkeit von Lehrerfortbildung [Learning on the job – Empirical findings on the effectiveness of in-service teacher education], in Müller F. H., Eichenberger A., Lüders M. and Mayr J. (ed.), *Lehrerinnen und Lehrer lernen – Konzepte und Befunde zur Lehrerfortbildung*, Münster: Waxmann, pp. 51–70.

Loucks-Horsley S., Hewson P. W., Love N. and Stiles K., (1998), *Designing Professional Development for Teachers of Science and Mathematics*, Thousand Oaks: Corwin.

Loughran J., Milroy P., Berry A., Gunstone R. and Mulhall P., (2001), Documenting science teachers' pedagogical content knowledge through PaP-eRs, *Res. Sci. Educ.*, **31**, 289–307.

Magnusson S., Krajcik J. and Borko H., (1999), Nature, source, and development of pedagogical content knowledge, in Gess-Newsome J. and Lederman N. G. (ed.), *Examining Pedagogical Content Knowledge*, Dordrecht: Kluwer, pp. 95–132.

Mamlok-Naaman R., Ben-Zvi R., Hofstein A., Menis J. and Erduran S., (2005), Influencing students' attitudes towards science by exposing them to a historical approach, *Int. J. Sci. Math. Educ.*, **3**, 485–507.

Mamlok-Naaman R. and Eilks I., (2012), Action research to promote chemistry teachers' professional development – cases and experiences from Israel and Germany, *Int. J. Sci. Math. Educ.*, **10**, 581–610.

Marx R., Freeman J., Krajcik J. and Blumenfeld P., (1998), Professional development of science teachers, in Fraser B. J. and Tobin K. G. (ed.), *International Handbook of Science Education*, Dordrecht: Kluwer, pp. 667–680.

Matthews M., (1994), *Science Teaching: The Role of History and Philosophy of Science*, New York: Routledge.

McCloskey M., (1983), Naive theories of motion, in Gentner D. and Stevans A. L. (ed.), *Mental Models*, Hillsdale: Lawrence Erlbaum, pp. 299–324.

Project 2061, (1989), *Science for All Americans*, Washington: AAAS.

Richardson V. and Placier P., (2001), Teacher change, in Richardson V. (ed.), *Handbook of Research on Teaching*, Washington: AERA, pp. 905–947.

Shulman L. S., (1986), Those who understand: knowledge growth in teaching, *Educ. Res.*, **15**(2), 4–14.

Shulman L. S., (1987), Knowledge and teaching: foundations of the new reform, *Harvard Educ. Rev.*, **57**, 1–22.

Sprinthall N. A., Reiman A. J. and Thies-Sprinthall L., (1996), Teacher professional development, in Sikula J., Buttery T. J. and Guyton E. (ed.), *Handbook of Research on Teacher Education*, New York: Macmillan, pp. 666–703.

Taitelbaum D., Mamlok-Naaman R., Carmeli M. and Hofstein A., (2008), Evidence-based continuous professional development (CPD) in the inquiry chemistry laboratory (ICL), *Int. J. Sci. Educ.*, **30**, 593–617.

Tobin K., Tippins D. J. and Gallard A. J., (1994), Research on instructional strategies for science teaching, in Gabel D. L. (ed.), *Handbook of Research on Science Teaching and Learning*, New York: Macmillan, pp. 45–93.

Van Driel J. H. and Justi R., (2006), The use of the Interconnected Model of Teacher Professional Growth for understanding the development of science teachers' knowledge on models and modelling, *Teaching and Teacher Education*, **22**, 437–450.

Van Driel J. H., Verloop N. and de Vos W., (1998), Developing science teachers' pedagogical content knowledge, *J. Res. Sci. Teach.*, **35**, 673–695.

CHAPTER 3

Top-down Approaches for Chemistry Teachers' In-service Professional Development – From Basic to Advanced

The literature generally suggests that the most promising professional development for teachers is long-term, continuous, and bottom-up (see Chapter 1 of this book). The combination of these three components allows teachers to become familiar with the intended change and develop a sense of ownership. However, suitable long-term and bottom-up programs are not always available, particularly in remote areas. A solution for professional development in these areas is the establishment of small networks of teachers within their schools or within the regional neighborhood. Local and regional networks provide teachers with opportunities to learn from each other and to jointly develop their teaching practices. However, unaccompanied local networks hold the risk of perpetuating existing practices, instead of developing new approaches and innovations. For new approaches to be adopted, input from educational research and curriculum innovation is usually necessary. As noted in Chapter 2, access to information, materials, and experts from outside the local environment is required. Traditional ways of accessing information include journals, books, meetings, and so on. Digital communication has extended these traditional possibilities through the sharing and exchange of information *via* the internet and social media, among others. The same holds true for chemistry teachers' professional development. This chapter discusses potential ways, from more basic to advanced, of providing information from outside the teacher's own school environment.

Advances in Chemistry Education Series No. 1
Professional Development of Chemistry Teachers: Theory and Practice
By Rachel Mamlok-Naaman, Ingo Eilks, George Bodner and Avi Hofstein
© Rachel Mamlok-Naaman, Ingo Eilks, George Bodner and Avi Hofstein 2022
Published by the Royal Society of Chemistry, www.rsc.org

3.1   Information Resources for Chemistry Teachers

3.1.1   Traditional Media

Traditional means of continuing chemistry teachers' professional development include the regular reading of domain-specific publications in their field. These are handbooks and materials on the teaching and learning of chemistry, and journals whose purpose is to inform teachers about chemistry research and chemistry education research, new approaches to teaching chemistry, and new materials available for chemistry education. Such journals are often published in either national or local languages and are specific to the national or local context of the teaching practice.

Journals for chemistry teachers and chemistry teacher educators can follow different approaches. In addition to journals that communicate new frontiers of chemistry research to the public and teachers, specific chemistry education journals report the results of chemistry education research. Some of these journals follow the format of traditional scientific journals; others more closely resemble magazines. In general, there are four different types of journals:

- Pure research journals, such as *Chemistry Education Research and Practice* published by the Royal Society of Chemistry in the UK. In general, the articles are comprehensive, relatively long, and well referenced to the prior literature. Journals of this type are mostly available in English and a few other major languages.
- Journals that belong to the category of research but provide shorter and more practical articles, *e.g.*, about new school-type experiments or lecture demonstrations. One example is the *Journal of College Science Teaching*. The *Journal of Chemical Education* published by the American Chemical Society (ACS) combines some of the features of the first category as well as articles that would fall into this second one. Journals of this type are frequently available in languages other than English.
- Journals focusing on the needs of practicing teachers, which report ideas for making changes in the way chemistry is taught. Articles in these journals are often written by teachers for teachers. Typical examples in the English language are *School Science Review* from the UK, *Chemistry in Action* from Ireland, *Chem 13 News* published in Canada, and the new *Chemistry Solutions* – an online periodical prepared by the American Association of Chemistry Teachers (AACT) in the USA. Typically, these journals cover short, easily digestible articles that are published in the local or national language.
- The recently emerged magazine-type journals, such as *Science in School, The Learning Teacher Magazine* or *Chemistry Solutions*. These journals provide access to new teaching ideas in a well-illustrated manner and are often distributed for free, either in print or online.

All of these magazines, as well as handbooks and materials provided by commercial textbook companies, national science teacher associations, and the educational divisions of the corresponding chemical society, can provide teachers with input to help them develop their own teaching practices, or form a basis for discussion among colleagues within the teacher's own school or across local and regional networks.

3.1.2 Online Resources

New methods of communication and digital media have vastly enriched the information pool accessible to chemistry teachers. Numerous websites deal with ideas for teaching and provide traditional and digital media content. Single sites offer worksheets for direct printing out; YouTube videos explain how to perform a particular experiment or activity; blogs and forums allow teachers to ask each other questions and discuss teaching ideas, and enable mentoring of novice teachers by more experienced ones. These new forms of digital communication offer chances to be better connected to information and colleagues, especially for teachers working in remote or isolated environments, where face-to-face interaction is not possible.

The new world of media not only offers different modes of communication, it also helps minimize the risk involved in professional changes by teachers, who can collaborate with their peers. The new media, however, has a fundamental problem. In traditional teacher materials issued by the chemical industry, chemical societies, science teacher associations, or commercial publishers, one can expect a reasonable standard for the validity of the published information; this is not always the case on the internet or with any social media. Teachers need to check very carefully whether the suggested teaching strategies are scientifically sound, are in line with modern educational theory, and are adaptable to their own classroom teaching practices. This can start with simple issues such as respecting the "prudent practice" in safety regulations that might differ from country to country. It might also be inconsistent with what the national or local school curriculum expects and suggests.

A potential problem with using materials from the internet is that they are not always correct. Some resources contain serious mistakes in content, while others use questionable models or illustrations. Teachers need to be highly reflective when choosing and adapting materials or teaching ideas from the Internet to avoid causing confusion and introducing misconceptions among their students (Eilks *et al.*, 2010).

3.2 Face-to-face Approaches for Chemistry Teachers' Professional Development

3.2.1 Teacher Conferences

Another form of traditional ongoing professional development for teachers, and chemistry teachers in particular, are conferences that are generally organized by national or regional science teacher associations or the educational divisions of the corresponding chemical societies. In addition, there are international conferences for chemistry teachers, such as the Biennial Conference on Chemical Education (BCCE) and the International Conference on Chemical Education (ICCE), having a long history of being organized by the ACS Division of Chemical Education and the IUPAC, respectively. In recent years, conferences for teachers have begun to be operated on an international level in Europe as part of educational policy by the EU. Many professional development projects under the auspices of the Seventh Framework Programme of the EU have offered conferences for teachers to familiarize them with the different curriculum innovation projects and their outcomes. Finally, a series of teacher conferences are now being offered under the scheme of *Science on Stage*, the *Hands-on-Science Network* and other initiatives.

Participation in teacher conferences provides more than just an opportunity to "see and hear" new ideas for teaching. It also provides a good place to test new hands-on experiences and to discuss them with colleagues who might come from different educational environments.

3.2.2 Continuing Professional Development (CPD) Workshops for Teachers

A common format for traditional chemistry teacher professional development consists of full- or half-day workshops offered by external professional development providers. An example of this would be the chemistry teacher professional development centers in Germany called *GDCh–Chemielehrerfortbildungszentrum*. These are regional hubs for chemistry teacher in-service training that are funded and supported by the German Chemical Society (GDCh). Today, there are seven of these centers across Germany, in Bremen/Oldenburg, Dortmund, Frankfurt, Karlsruhe, Leipzig/Jena, Nuremberg, and Rostock. These in-service training centers are situated at different universities, each serving a certain region. Each center is connected to one or two academic chemistry education groups. The centers offer half- and full-day courses on modern developments in chemistry, new topics and experiments, innovative pedagogies, or the transfer of chemistry education research findings into classroom practice. Most of the centers offer courses both at the university and in schools throughout the corresponding region. Typical events start with a lecture on an innovative topic, *e.g.*, the use of bioplastics, biofuels, or nanomaterials (see Chapter 8 of this book). The presentation is followed by a workshop or laboratory session where the teachers can try out experiments or activities and then discuss how to adapt them to their own classrooms.

Although these professional development workshops offer an opportunity to try out new activities, they are limited in their potential for initiating long-term change. Not much is known about the effectiveness of such single-contact events. For many teachers, however, this is one of the few chances they have to try out new experiments or experimental techniques, or to obtain hands-on experience in the use of new media and equipment under the guidance of experts.

3.3 Challenges and Resources for Chemistry Teachers' Professional Development – A View from the USA

Advice on the design of professional development programs for chemistry teachers in the USA is readily available, *e.g.*, Loucks-Horsley *et al.* (2009), or the *Journal of Science Teacher Education*, whose mission is described on the journal's website as "...disseminating high quality research and theoretical position papers concerning preservice and inservice education of science teachers." A segment of the literature on professional development that might be particularly useful consists of reports of research on factors that make these programs successful (Wade, 1985; Guskey, 2003). A special resource is provided by the results of the Salish project funded by the US Department of Education between 1993 and 1997 (Brunkhorst *et al.*, 1993; Adams and Krockover, 1997; Simmons *et al.*, 1999; Yager and Simmons, 2013). The Salish project has been described as "...an exploratory study conducted to uncover knowledge about the relationship between secondary science and mathematics teacher preparation, new teacher knowledge, beliefs, and performances; and student learning outcomes" (Yager and Simmons, 2013).

Traditionally, professional development by well-intentioned chemists and educators in the USA has been based on a philosophy of doing things to, or possibly for, teachers. Today, it is more often done with teachers. There seems to be a tacit assumption behind the development of many professional development programs that individual teachers or groups of teachers cannot pursue professional development on their own. It can only occur when teachers enroll in courses or attend workshops run by people who may have been teachers at some point in their careers, but are now faculty in education programs or chemistry departments.

Chemistry teacher professional development in the USA is complicated by the magnitude of the problem. In the fall of 2017, more than 50 million students were enrolled in close to 100 000 public senior high schools that employed 3.2 million teachers (US Department of Education, 2016). Historically, the American Chemical Society (ACS) focused on the more than 30 000 chemistry teachers working with the 15 million students enrolled in grades 9 through 12, each year. Today, the ACS is developing instructional material and teacher networks that can also reach some of the more than 75 000 middle-school teachers who teach physical sciences in the USA, and is continuing efforts to develop activities that can be used in elementary school classrooms as well.

The need for professional development of high-school chemistry teachers is greater today than ever before. According to data from the National Center for Education Statistics, enrollment in high-school chemistry courses in the USA has increased from 32% in 1982 to 49% in 1990 and 70% in 2009 (US Department of Education, 2016). Part of this change can be attributed to the development of the curriculum *Chemistry in the Community* by the ACS or, as it is commonly known, ChemCom (ACS, 1988). The goal of ChemCom is to provide a high-school chemistry course that covers chemistry concepts in the context of societal issues, using real-world examples.

ChemCom deserves mention in this book on professional development because the ACS did something for this textbook that many other curriculum development projects have failed to do. It took on the task of creating a longstanding series of professional development workshops to help teachers understand how to implement this new approach to teaching chemistry. Thirty years after the first edition appeared, a series of *Chemistry in the Community* webinars is still available on the ACS Education Office's website. One of these webinars introduces the rationale behind the ChemCom curriculum, whereas others describe the teaching philosophy and content of individual chapters.

As the world's largest scientific organization, with more than 150 000 members, the ACS has the volunteer resources needed to prepare a variety of materials that can, and often are, incorporated into professional development. The professional development also covers an ACS Education Office website presenting information on safety in both middle and senior high-school classrooms. These include links to safety data sheets (SDS), chemical safety for teachers, guidelines and recommendations for teaching high-school chemistry, chemical health and safety resources, safety guidelines for community activities, and safety tips for children in both English and Spanish.

There is abundant content for the professional development of high-school chemistry teachers on the ACS Education Office website. It includes the *ChemMatters* magazine, which looks at chemistry in everyday life, ideas for introducing the core concept of energy, lesson plans based on the ACS National Historic Chemical Landmarks, Green Chemistry teaching resources, and a link to the ACS examinations as a source of high-quality student assessment and study materials. This website has links to both the ACS Guidelines and Recommendations for the Teaching of High School Chemistry and High School Chemistry in the National Science Education Standards. As might be expected, this website has a section entitled Professional Development, which describes workshops and seminars for chemistry and science teachers.

Efforts to address the professional development needs of high-school chemistry teachers in the USA can be traced back to 1924, when the *Journal of Chemical Education* was launched by the newly formed ACS "Section of Chemical Education" (Bodner, 2011), which eventually became the ACS Division of Chemical Education, with more than 5000 members. In 1987, the

ACS began publishing *WonderScience*, an eight-page magazine issued four times a year to introduce activities that children and adults could do together. Two years later, the American Institute of Physics began to co-sponsor *WonderScience*, which was then published eight times a year. In 1997, two volumes of the Best of Wonder Science appeared that are still available on the ACS Education Office website. Intended for use with grades 3–6, they represent the ACS's first step toward developing materials for elementary school science.

The commitment of the ACS to providing materials for senior high-school instruction has taken another step forward based on the philosophy of "teaching science through doing science", which has involved the development of materials aligned with the *Next Generation Science Standards* (National Research Council, 2013) for grades 6–8 that can be accessed at www.middleschoolchemistry.com. This site contains sample lesson plans, multimedia, and a list of workshops for chemistry teachers held at National Science Teacher Association (NSTA) meetings. For many years, there has been a close working arrangement between the NSTA and the ACS. The NSTA asked the ACS to help increase chemistry content by incorporating a "day of chemistry" that was both content-rich and pedagogy-rich to be held at three of the annual NSTA meetings. The goal of this collaboration was to introduce content that allowed teachers to "grow and stretch".

Several years ago, the ACS Board of Directors authorized the creation of a new organization known as the American Association of Chemistry Teachers (AACT) (Bodner, 2014a,b). One of the metaphors developed for this new organization has already been introduced in this section. The AACT is not a continuation of attempts by the parent organization to do things either to teachers or for teachers, but rather with teachers. The resources of the ACS were made available to create an organization run by high-school teachers, for high-school teachers. The AACT reflects the belief that chemistry content can, and should, appear in elementary- and middle-school courses, as well as in the traditional high-school chemistry courses. The philosophy of the AACT is captured on their website as "Share. Connect. Succeed", and it is based on the assumption that professional development materials can be used by individual teachers to shape their own classrooms, by groups of teachers working together, as part of pre-service instruction, and as the source of materials for professional development programs.

In addition, there are many local and regional initiatives, *e.g.*, by the faculty and staff at Purdue University in Indiana. Many professional development workshops for teachers from the primary grades up through the advanced high-school chemistry course are regularly provided. As one example, the Purdue University chemistry education research group was involved in a large-scale, National Science Foundation (NSF)-funded National Center for Learning and Teaching (NCLT) in Nanoscale Science and Engineering Education (NSEE). Whereas others had tried to imagine what activities would be good for introducing NSEE to senior high-school students, we started by asking: "What kinds of activities interest middle- and high-school students?" Whereas conventional wisdom among members of the NCLT assumed that students would be interested in activities related to "everyday life", we found that these students were unusually interested in topics such as a gecko's ability to climb walls, which is far from an everyday experience in the USA. We concluded that a more appropriate term was "real-world example". We also noted that these students were interested in activities for which they knew something about the science but did not feel that they "understood" it (Hutchinson *et al.*, 2011).

Traditional professional development for in-service teachers has focused on enhancing teachers' content knowledge (CK), pedagogical knowledge (PK), or pedagogical content knowledge (PCK) that builds on prior knowledge or experience. Nanoscale science and engineering, however, presented a unique challenge for teacher education workshops because of its unfamiliar content (Wischow *et al.*, 2013). Our study concluded that teachers' implementation of nanoscience was influenced by a combination of teaching orientations, CK, knowledge of students, and knowledge of pedagogical strategies. We noted that the challenge of introducing new content provided a context around which teachers could think about both new ways of teaching and restructuring the content that they taught.

When the term professional development is coupled to chemistry it usually brings to mind images of workshops or other activities with, or for, middle- or high-school teachers. There is, however, also professional development within the context of college or university chemistry faculty. A look into practices in the USA might give us an idea of the potential focus of these corresponding initiatives:

- In much the same way that the ACS has had a longstanding commitment to professional development of high-school teachers implementing the high-school curriculum known as ChemCom, it also provides hands-on professional development for faculty implementing its curriculum *Chemistry in Context* (ACS, 2017) for use in a one-semester college-level course for non-majors.
- *Chemistry Collaborations. Workshops & Communities of Scholars* (cCWCS) is the outgrowth of a project supported since 2000 by the NSF that has sponsored more than 100 five-day workshops for faculty engaged in undergraduate teaching. In 2017, this group offered workshops on such diverse topics as chemistry in art, distance learning and hybrid teaching, active learning in organic chemistry, materials science and nanotechnology, and forensic science. The cCWCS was involved in recruiting participants for the *Advancing Chemistry by Enhancing Learning in the Laboratory* (ACELL) project, described in the following, when it was brought to the USA.
- Having worked with colleagues from Australia, Orgill and Bodner (2014) brought the ACELL model to the USA. ACELL focuses on helping faculty work with students from their institution to bring new perspectives to the laboratory course associated with the introductory, college-level

general chemistry course. Buntine *et al.* (2007) described ACELL as a multi-institution collaborative project involving chemistry faculty from across Australia. The ACELL project brought together faculty who taught undergraduate chemistry courses "to improve the quality of learning in undergraduate laboratories through two interlocking mechanisms". One of their goals was to develop a database of educationally sound experiments that "worked". However, the other goal was to create a viable mechanism for professional development of both faculty and students. Faculty who participated in this workshop had to be accompanied by an undergraduate from their department and had to submit an example of a laboratory from the course that they taught before attending the workshop. Faculty and students worked together to "perform" experiments of different types and each day's experiences were discussed that evening in an informal gathering.

3.4 Top-down, Long-term Approaches for Chemistry Teachers' Professional Development – Three Cases from Israel

3.4.1 Preparing Biology and Agriculture Teachers to Become Chemistry Teachers – A Case from the Upper Galilee in Israel

A preliminary survey of the teaching of science and technology conducted in the Upper Galilee in Israel in 1989 indicated that there was a severe shortage of chemistry teachers in that area. As a result, chemistry was only taught in a limited and marginal manner in secondary schools in the northern district of Israel. Following this survey, it was decided to open a course to prepare biology and agriculture teachers to teach chemistry.

The teachers who participated in the course were experienced and had an academic degree in the science that they taught. Most of them had appropriate pedagogical training to teach their subjects, but they lacked the ability to teach chemistry, especially at the upper secondary-school level. The designers of the course assumed that a multistage program including content in chemistry and the special teaching methods for this subject would prepare these teachers properly to teach chemistry.

The program was implemented between 1989 and 1991. The teachers met each week for eight academic hours (224 academic hours per year). The program consisted of intensive coursework that included teaching the content of the high-school chemistry curriculum at the basic and advanced levels, activities, diversifying teaching methods and pedagogical strategies.

The teachers who participated had more than 10 years of teaching experience in high schools in various science subjects, and therefore the course coordinators did not have to deal with problems that often occur with novice teachers. Lortie (1975) had claimed, however, that if teachers do not start their work gradually, the little experience they have accumulated during the training period does not allow them to function properly in schools. Teachers often face problems that they are not sufficiently familiar with, such as lack of motivation among students, difficulties in understanding, class heterogeneity (Hall and Loucks, 1981), or poor discipline. Teachers must be equipped with content in both the sciences and teaching methods, *e.g.*, PK in the field in which they are engaged (Constable and Long, 1991).

The goals of the course were:

- to impart the appropriate CK for the chemistry course contents in Israeli high schools and to present the curriculum for high schools according to the syllabus set by the National Committee of Chemistry Education in Israel
- to introduce PCK (Shulman, 1986) associated with: (i) awareness of learning difficulties in chemistry and their treatment, (ii) integration of the relevance of chemistry in everyday life, (iii) diversity of learning methods while recognizing the heterogeneity of the student group, and (iv) development of learning materials on various topics, lesson plans on integrative subjects, or enrichment materials for teachers
- to support teachers in the early stages of classroom instruction
- to create a core of "expert" teachers who could be "leading teachers", who would then guide their colleagues and pass on to them the knowledge and teaching methods and skills they had learned for solving students' learning problems.

A total of 42 teachers participated in the course. About a quarter of these teachers continued to participate for 4 or 5 years. Eight teachers did not continue their participation after the first year because they were unable, for various reasons, to teach chemistry in their schools. A number of teachers also left each year for personal reasons. In the last year, however, a number of experienced chemistry teachers, most of them new immigrants, joined the course, participating in activities that marked this year, as will be described below.

The course could be divided into three categories:

- Imparting the content of the subject (in the first stage, within the scope of the basic chemistry curriculum for high-school students, and then in an extended scope).
- Introducing aspects related to the operation of the program in classrooms, such as visiting industrial chemical plants, teacher training, and the promotion of teacher leaders in their schools and in the region.
- Demonstrating the practical aspects of teaching chemistry and its relevance to daily life.

The various activities involved discussions related to learning difficulties and misconceptions that occur when teaching certain concepts in chemistry.

These activities helped the teachers assimilate new concepts in chemistry. They also helped them understand students' perceptions, and the difficulties they face while solving problems in chemistry, namely:

- Identifying students' learning difficulties in the field of chemistry studies, and finding ways of teaching to overcome these difficulties (Ben-Zvi *et al.*, 1986, 1987).
- Examining the causes of learning difficulties based on the theories of Piaget and Oswald (Novak *et al.*, 1971).
- Understanding the way in which students learn in chemistry with respect to Piaget's theory of thought development (Herron, 1975; Gabel and Sherwood, 1981).

Learning groups usually consist of a heterogeneous population of students – in both their ability to learn and their motivation for learning (Hofstein and Kempa, 1985). Within the context of motivation and personality, Adar (1970) categorizes students according to the characteristics of achievement, curiosity, and conscience. Social understanding of the behavior of each of these student attributes contributes to understanding the relationship between their learning and their achievements. To motivate students to learn and invest effort, Kolesnick (1978) suggested varying instruction and choosing instructional strategies so that each student would be interested and able to participate in the learning process. Using a variety of method types increases the chances of reaching most students (Eilks *et al.*, 2013).

The enrichment part of the course included issues such as: chemistry in everyday life, industrial chemical plants, nature of chemistry, and historical background of chemistry. The topics were also varied, *e.g.*, oil, Dead Sea plants, fertilizers and cosmetics, radioactivity, and air pollution. The teachers had to choose one topic according to their ability and interest, and develop an enrichment learning unit for the high-school students (approximately 20 periods).

Critical reading of articles related to science, taken from scientific journals or daily newspapers, was suggested as a strategy for diversifying content or teaching strategies. This is an informal type of learning, which allows for an extension of learned subjects, and an updating of knowledge on current issues, acquisition of critical skills, and selection of important concepts (Wellington, 1991). Through the daily press, for example, the teachers were able to search for content related to energy sources or oil prices and try to analyze various scientific issues socially, economically, and politically (see Chapter 8 of this book). Sometimes the subject became interdisciplinary, and the distinction between fields of science was blurred. In this case, teachers and students were able to learn how difficult it is to separate the various fields of science, technology, and engineering, and how closely related they are to each other.

The articles were adapted to the students' level of knowledge and ability. The teacher learned to ensure that the article would not be too long and that it included a proper balance between new and familiar concepts. It was sometimes necessary to rewrite the article or to prepare guide questions to help students understand it, because many students did not distinguish between the main point of the treatment and the difficulty in reading comprehension. The teachers also learned to ask the students to prepare guide questions, or to highlight key phrases or keywords. Thus, the students would be employed, individually or in groups, in various methods, such as group investigation methods (Sharan and Hertz-Lazarowitz, 1980).

Even in the first year, the teachers felt responsible for the success of the course, sharing programmatic or methodological problems, ideas, and lesson plans with each other. At their own initiative, they came to various meetings with suggestions of exams, quizzes, work papers, and interesting experiments. Because they were asked to work on different tasks in the teams, each team became a kind of supportive subgroup that continued to operate independently of the course. The work of these teams was expressed primarily during the last year of the course, when participants were asked to develop learning materials on integrative subjects.

Evaluation was carried out mainly during the second year of the course within the framework of two main selected areas:

- contribution of the course to participating teachers
- the status of teaching chemistry in the Upper Galilee region.

Data were collected by means of questionnaires given to the participating teachers, interviews with 10 of them, telephone conversations with some of the teachers who dropped out of the course, and questionnaires to the schools whose teachers participated in the course.

The evaluation of the contribution of the course to teachers' development was based mainly on the final questionnaire completed at the end of the year by all of the participants, the interviews with 10 of them and the informal conversations held during group visits and meetings. Various components of the course were addressed, including the self-efficacy to teach the curriculum subjects, teaching methods, computer integration into teaching, and experience in writing learning materials. In the final questionnaire, the teachers were asked to indicate to what extent the course contributed to strengthening and expanding their knowledge of a number of key topics and concepts. The participants were also asked about other activities related to teaching methods and about the contribution they received on topics related to teaching methods, such as developing models and teaching methods, constructing tests, and integrating computers into chemistry teaching.

An examination of the teachers' answers indicated an appropriate contribution to the development of different models and teaching methods and a smaller contribution to the topics of constructing tests and incorporating computers into their teaching. In general, and in the structure of this course in particular, the detachment of methodology, methods of instruction, and teaching strategies from the subjects themselves was essentially artificial. In the interviews, most of the teachers mentioned their extensive use of the materials that they received during the course (work papers, articles, and other enrichment materials) and the new ideas that were drawn from the meetings. The following quotes provide two examples of what was mentioned during the interviews:

> The course gives a lot in terms of methodology, reference materials and enrichment, really the methodology of the subjects.

> Around the basic core [of the subject] was the growing radius of a lot of aids, and a lot of things that came later from the course itself, from the people—work pages, tests, or articles.

During the interviews, the teachers also mentioned the substantial benefits they gained from the course in terms of familiarity with and control of various subjects. A new immigrant teacher, for example, who was a veteran chemistry teacher said:

> I did not know the subject of the bromine industry. In Russia, the processes are completely different, and I had to take the course on the bromine industry. Sugars, to tell you the truth, I did not even know what sugars were, where to start, what models, *etc.*, *etc.*.... In this course I also heard about proteins. We heard here about all of the subjects that we teach at school...

Another chemistry teacher, who joined the course in the last year, said:

> I joined with apprehension because I knew it was a chemistry course for biology teachers and I thought it was not for me. The subjects of sugars, proteins, polymers, I have never taught, and this I got here.

For the purpose of working in groups, the participants formed six teams, each with two to four teachers, except for one group consisting of five teachers. Most of the team members met an average of eight times, but some met only twice, and at least some of the meetings were held in the afternoons and evenings in the teachers' homes. The teachers' reactions to this activity indicated that the practice of writing in the teams contributed to their professional development in several ways:

- Recognizing the advantages of teamwork. Some teachers noted that developing a particular topic in group work was mutually productive, efficient in terms of time utilization, and resulted in high-quality products.

- Broadening the horizons of subjects that are not part of the curriculum. Many teachers mentioned that they spent a lot of time searching for new material, which provided an opportunity to get to know new topics and bring up new ideas, mainly in methodological contexts.
- Practice in writing a subject and adapting to goals and teaching methods. For some of the teachers, writing about CK and teaching methods was a new experience that added another aspect to the diversity of everyday work.
- Strengthening social relations. Working in small groups contributed to strengthening personal relationships between participants.

Because the main goal of the program was to equip the teachers involved in the project with chemistry teaching tools and to provide them with confidence in their ability to teach chemistry, their feelings were examined from this perspective. In general, they saw themselves as chemistry teachers and were confident in their ability to prepare students for the final examinations that were designed and administered centrally by the Ministry of Education, at least at the basic level.

The confidence that they demonstrated and their "pride in the unit" relied on two components:

- The quality of the training they received and the control they acquired of the curriculum content, teaching methods, ideas for enrichment and more.
- The direct contact with the course coordinators and the deep feeling that they would always be able to apply for help and advice on any subject and would receive an answer.

During the second year of the course, the teachers started to teach chemistry in their schools, and therefore needed support. The course coordinators visited the schools every 2 weeks and observed the lessons given by the course participants. After each lesson, a discussion was held on the teaching methods, the problems raised by the students about the subject being studied, and the learning difficulties that were encountered. Other chemistry teachers who did not take the course and taught at the same schools joined the discussion.

Workshops with the program's teachers were sometimes held in the schools (about five workshops a year). These workshops were based on sample lessons created by the course participants. This activity included analysis of the lesson, alternative proposals for teaching the content, recommendations for additional experiments or material for enrichment, and the preparation of supplementary material for students.

In conclusion, we noted that the teachers expressed great satisfaction with the course and showed confidence in their ability to teach chemistry to maturity. The teachers, who also taught biology, reported a positive impact of the course on teaching that subject as well. A review of the structure and

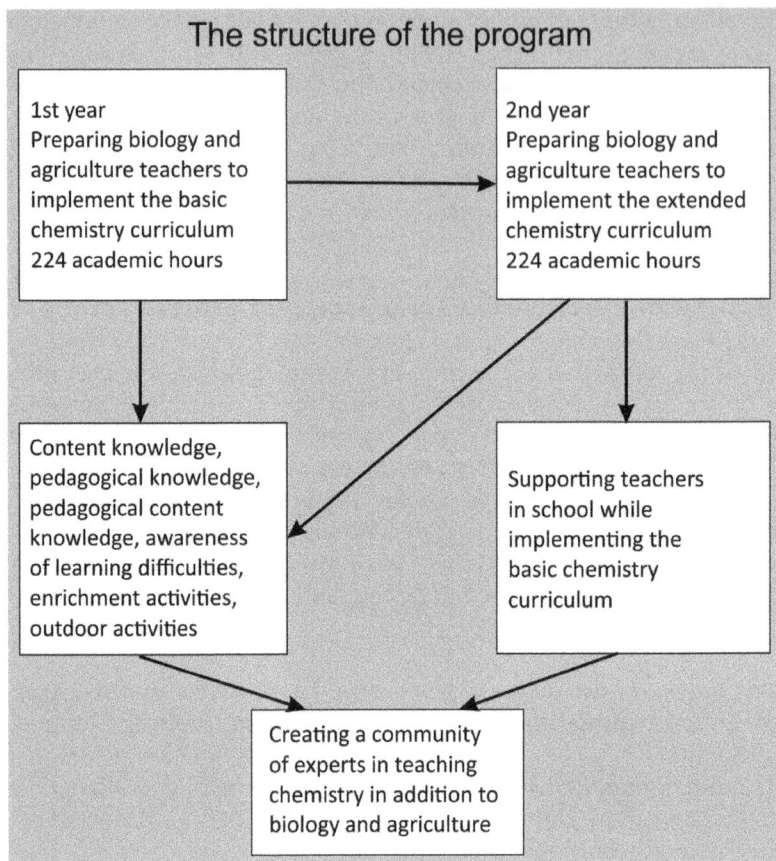

Figure 3.1 Structure of the CPD program.

contents of the course indicated a gradual, repetitive, spiral progression: teaching subjects to teachers, supporting and guiding teachers during their teaching, imparting new subjects to teachers, enriching and deepening previous subjects, and repeating the whole process. The structure of the course had an element of dynamic development, which apparently contributed to its success. Figure 3.1 summarizes the structure of the program.

3.4.2 Updating Chemistry Teachers' CK and PCK on Current Issues of Chemistry in a Long-term CPD Program

In their article *Teacher development as personal, professional, and social development,* Bell and Gilbert (1994) wrote that effective professional development for teachers should include the development of both CK and PCK. The third component of Bell and Gilbert's suggestion was the social dimension, which involves learning new ways to work with other people in the educational system. They argued that these three dimensions are interrelated and that the development of one aspect cannot proceed unless the other aspects improve as well.

With respect to learning materials, Kempa (1983) noted that chemistry should include the following dimensions: the conceptual structure of chemistry, the processes of chemistry, the technological manifestations of chemistry, chemistry as a "personally relevant" subject, the cultural aspects of chemistry and finally, the societal implications of chemistry. More specifically, it was suggested that in the teaching and learning of chemistry, students should be exposed to recent investigations. Moreover, chemistry should be viewed as an inquiry-based discipline, giving rise to new knowledge and insights. To this end, problems can be solved in both the classroom and the laboratory using inquiry-type activities and methods. This approach enables the students to ask questions, plan and conduct investigations, think critically, construct and analyze alternative explanations, and express scientific arguments (Bybee, 1997). In addition, to make it more relevant to students' lives and to the society in which they live and operate, chemistry should be taught as an applied science of major economic and technological importance (Stuckey *et al.*, 2013; and see Chapter 8 of this book).

Clearly, such an approach to teaching chemistry puts great demands on chemistry teachers. Traditionally, in many countries, chemistry teachers in their pre-service preparation are exposed primarily to the CK of chemistry, *i.e.*, the conceptual structure and processes of chemistry (see Chapter 1 of this book). Modern research in chemistry, its technological applications, its influence on society, and its cultural characteristics are neglected or receive only limited attention, with consequences for the functional chemistry curriculum (Hofstein *et al.*, 2011). However, teachers are considered to be the central element in reforms in teaching chemistry that ultimately influence numerous students, and they therefore need sufficient opportunities to achieve life-long learning on the current developments of modern chemistry. This point of view is supported by an international educational policy document by Osborne and Dillon (2008), as well as other reports (European Commission, 2004, 2007; Dillon and Osborne, 2007). The reports all reflect a consensus on the importance of good-quality teachers, regularly updating their knowledge base:

> Good quality teachers with up-to-date knowledge and skills are the foundation of any system of formal science education. Systems to ensure the recruitment, retention, and continuous professional training of those individuals must be a policy priority in Europe. (Osborne and Dillon, 2008, p. 25)

Research findings on the effectiveness and professional development of teachers often underscore the importance of teachers' CK and professional interest, as well as their PK (Munby *et al.*, 2001). What teachers know and how the connection of CK and PK in chemistry differs from other subject areas were captured by Shulman (1986) in his idea of PCK. Shulman (1986) described PCK as knowledge, "which goes beyond knowledge of the subject matter...to the dimension of subject matter knowledge for teaching" (p. 9, and see Chapter 2 of this book).

To develop the necessary PCK, teachers usually attend relevant seminars during their pre-service teacher education program, and they then have to update their knowledge by participating in professional development programs initiated at their schools or by external providers of professional development (Krajcik *et al.*, 2001). These professional development programs guide and support the teachers in better suiting their teaching to the needs of their students or in implementing a new curriculum (Loucks-Horsley and Matsumoto, 1999). Sometimes, however, these programs focus more on the teachers' PCK, and neglect issues of updating the teachers' CK (Kind, 2009).

Based on the above arguments, a special model for enhancing chemistry teachers' CK aligned with the corresponding PCK was developed at the Weizmann Institute of Science in the 2009–2010 academic year. The purpose of the program was to give science teachers the opportunity to update and enhance their CK of contemporary research such that they would be able to use it to reach their students (Mamlok-Naaman *et al.*, 2010). The program was intended to empower teachers and to increase their motivation by providing them with opportunities to develop professionally and to be involved in innovative activities, *e.g.*, visiting academic research laboratories, attending scientific workshops and seminars that deal with frontiers in science, and coming into contact with new developments in science education. During this program, it was anticipated that the teachers would gain some of the necessary CK that would enable them to successfully guide their students into new areas and developments in modern chemistry. This program was designed to initiate a meaningful and profound change in the curriculum and the teachers' teaching style and enhance their professional development, as well as their knowledge of teaching and pedagogy. CPD in this form was expected to broaden teachers' PK and CK.

The program, started at the beginning of the 2009–2010 academic year, was carried out collaboratively between science educators from the Department of Science Teaching and representatives of the different science faculties, among them the faculty of chemistry (at the Weizmann Institute). It involved teamwork and collaboration among three groups: scientists, science educators, and teachers. Moreover, the teachers were expected to participate in both scientific research and science teaching. It was assumed that the teachers, in collaboration with chemists and chemistry educators, would initiate and participate in innovative activities to advance chemistry education in Israel.

In general, the teachers' pre-service training program takes 3 years to complete. For the first 2 years, the teachers spend 2 days per week at the Institute and are remunerated with half of a MSc fellowship. In the third year, they spend 1 day a week at the Institute, their day off from teaching. In this way, the teachers are allowed to continue teaching and to engage in innovative school-based activities in parallel with their study. Most of the courses are given during the first 2 years of the program, whereas the third year is mostly devoted to seminar work and school-based initiatives. The science teaching courses for the different groups of participants (teachers of biology, chemistry, mathematics, and physics) differ, although there is one general course in which all of the teachers participate, *i.e.*, assessment in science teaching. The courses were specifically designed for the teachers and focused on three aspects: advancing central and contemporary topics in their discipline, enhancing school-related knowledge, and providing general introductory courses to serve as a foundation for other disciplines (*e.g.*, biology for chemists and physicists).

In the chemistry-related courses, a specific course design was applied, consisting of three steps:

- Course lectures together with the traditional MSc students. The teachers were therefore part of a large passive audience (including regular graduate students). These lectures were neither aligned with the school curriculum nor with teachers' ability to teach the content to their high-school students.
- A "follow-up" tutoring lesson, which was especially prepared for the teachers by one of the staff scientists, was targeted at elaborating upon the course lecture. This second step still focused on CK, but with more attention to pedagogy. The tutor paid more attention to the teachers' needs, and supported them in their ability to understand the content.
- A workshop coordinated by a researcher from the chemistry group of the Department of Science Teaching, to apply the scientific knowledge to the field of chemistry education. The third step focused on PCK. The teachers used their CK and PK to transform them into PCK.

The three steps differed from each other in the aspect of the knowledge provided to the teachers, as shown in Table 3.1.

Learning advanced chemistry is not an easy task for teachers who have often completed their formal education more than 10 years prior. Several factors in the teachers' background can inhibit learning. Those teachers that completed their studies many years ago generally do not remember the content of the basic courses that they took during their initial teacher education studies. Furthermore, scientific (chemistry) knowledge has greatly advanced since that time, and there is a consequent gap between the

Table 3.1   Levels of CK, PK, and PCK in the three steps (Mamlok-Naaman *et al.*, 2010).

Step	CK	PK	PCK
Step 1. *Scientific lecture*	High	Low	Low
Step 2. *Follow-up*	High	Medium	Low
Step 3. *Adaptation to education*	Medium	Medium	High

knowledge that the teachers originally learned and their knowledge of modern chemistry (Tuvi-Arad and Blonder, 2010). After the teachers finished their university chemistry studies for teacher education, they became professional chemistry teachers and focused on the high-school chemistry curriculum. Hence, for the program described here, advanced topics were chosen that were associated with the chemistry curriculum, both to overlap with the teachers' prior knowledge and to provide relevant content. In 2009–2010, two courses met these criteria: "Organic reactions used in the total synthesis of natural products" and "Spectroscopy". The teachers who participated in the program generally taught organic chemistry and therefore knew the underlying fundamental principles in this field. Spectroscopy is not taught as such, at the high-school level in Israel; however, spectroscopy is often integrated into other topics, such as environmental and physical chemistry. The following is a description of how the three-step design was incorporated into in the course "Organic reactions used in the total synthesis of natural products".

Step 1 – Lecture: An advanced course in organic chemistry was given that included advanced topics in organic synthesis, such as retrosynthetic analysis, C–H acidity, C–C bond formation, hydroboration, region-selective enolates, enamines, acyl anion equivalents, and rearrangements. The lectures provided oral explanations taught together with organic chemistry equations that the lecturer wrote on the blackboard. A written exercise was given after every lesson. The evaluation of the course was based on a test that consisted of questions similar to those given in the exercises.

Step 2 – Follow-up: A tutoring lesson was given after each lecture by a staff scientist or a graduate student in the organic chemistry group. This lesson was given separately to the teachers, while the chemistry MSc students had a different tutor. The follow-up lessons included three parts: the tutor answered teachers' questions and explained unclear issues that emerged after the lecture; the tutor then gave more examples pertaining to the material that was taught in the lecture; at the end of the lesson, the tutor introduced new material in organic chemistry that should help to support the next lecture. The structure of each lesson was flexible, and it was changed each time it was offered according to the teachers' needs. The tutor usually added more questions to the written exercise that was given by the lecturer.

Step 3 – Adaptation to education: This session was conducted by a staff member from the chemistry education research group. The focus was on applying the advanced chemistry content to the field of education, and using this CK to solve the exercises. Adaptation to education sessions was carried out using a format in which the teachers usually worked in pairs and an educational guide aided their learning. The assignment at this stage was to produce a poster that would help the teachers adapt their new CK to the content of the courses they teach.

The enactment of the program was followed by a study. Its findings revealed that the three steps supported the teachers in learning advanced scientific content and in making it part of their teaching repertoire, thus enhancing their PCK. We can conclude that, by and large, these goals were achieved. According to the teachers, however, there were many challenges that should be taken into account if this program is to be continued. During all three steps, the teachers repeatedly requested to reduce the course level and its demands, in part because the lectures and the reading material were in English. The participating teachers were quite intimidated by this difficulty at the beginning; however, they managed to cope with it and their English improved during the course. But when the individuals presenting the lectures changed, the teachers complained. These findings are consistent with arguments raised by Schleicher (2009) when referring to dimensions of challenge and support, to describe successful changes in educational systems (Brickhouse and Bodner, 1992).

The dimensions of challenge and support were carefully examined with respect to each of the three steps, which had a distinct role, structure, and learning environment. Although they are separate entities, each step depends on the other and they were designed to support each other. To promote success, different connections among the steps are needed, including a connection involving the content of the lectures and tutorial and a connection between the lecturers. The content of the tutorial should support the lecture's content without adding new content. The teachers expected the tutor to go over the material from the lecture with them. When the tutor introduced new content, the teachers did not have enough support and they had difficulties coping with the lecture.

Moreover, teachers experienced different relationships with the three people involved in the scientific course: the lecturer, the tutor, and the educational guide. The educational guide had the most open relationship with the teachers. The teachers felt free to ask any questions and even to complain. The educational guide was also present at the lectures and the tutorials, and therefore had a complete picture of the course. He/she was able to give the lecturers and tutors repeated feedback when necessary. It was crucial to the success of the program that the educational guide inform and guide the tutor regarding the teachers' needs. When the tutor was willing to receive this feedback from the educational guide, his/her tutorials supported the teachers' needs. The tutor also had to be in contact with the lecturer regarding the content of each lesson.

Another aspect that had to be carefully considered was the relevance of the scientific courses to the teachers' needs in their classes. As already mentioned, the selected courses were advanced topics associated, to some extent, with the high-school chemistry curriculum. The teachers immediately tried to apply the new content to their teaching. During the tutorials, they asked many questions that were aimed at clarifying issues they had regarding the curriculum, which resulted in deepening their understanding of the subject matter. In adapting the material to specific educational steps, the teachers were required to transfer the advanced knowledge they had learned in the course to teaching high-school students. They had to consider the students' knowledge, interests, and the curriculum in coming up with a product (a poster and a lecture for high-school students). During these sessions, where they worked in small groups, they frequently discussed the pedagogy of teaching specific content matter. This step gave the teachers an opportunity to enhance their PCK.

There was also evidence that the model contributed to the teachers' knowledge in each of three domains: CK, PCK, and the creation of a professional learning community (Wood, 2007). Based on the teachers' achievements in the course (examination and their improvement in the knowledge test), it is suggested that the approach contributed to teachers' CK. The analyses of the posters and the minutes showed that by mastering the CK and applying it to their requested assignment (the posters), the teachers also improved their PCK. Interestingly, the teachers experienced each step differently. They used the third step (adaptation to education), which was carried out as work in small groups (with guidance), to solve the exercises that were given by the lecturer and the tutor during the first two steps. They claimed that the three-step model supported their conceptual learning and enhanced their understanding of the course content. However, they were unable to connect the newly learned content to education until the last month of the course (Tuvi-Arad and Blonder, 2010). Nevertheless, when they managed to do so, they appreciated the contribution of the third step to their teaching profession and to their PCK.

A crucial contribution that emerged from the teachers' interviews was the creation of a "professional community of learners" (Mamlok-Naaman *et al.*, 2007). This occurred especially in the second and third steps, in which the teachers worked in small groups for two academic hours. The formation of such a community of learners contributed to their feeling of ownership (Mamlok-Naaman *et al.*, 2007), to their learning and to their teaching; as a result, it enhanced their PCK (Taitelbaum *et al.*, 2008). The teachers felt that they had gained more self-confidence to criticize their own work and to understand their teaching strategies, in addition to their feeling of ownership regarding the teaching of new up-to-date scientific topics.

3.4.3 Giving Chemistry Teachers New Ways to Move Toward Inquiry Teaching – The TEMI Project

Varying the classroom learning environment by implementing different types of instructional techniques (or pedagogical interventions) may have the potential to enhance students' situational interest and motivation in science, which are usually not triggered by traditional curricular and pedagogical approaches (Kempa and Diaz, 1990; Bolte *et al.*, 2013). Inquiry is suggested to be among the most promising ways to increase interest, motivation, and effective learning in science (Hofstein and Lunetta, 2004; European Commission, 2007; Abrahams, 2011).

Inquiry-based science education has the potential to elevate students' attainment levels and improve their attitudes toward science (Hofstein and

Lunetta, 2004; Lunetta *et al.*, 2007), but some students find it difficult to engage in inquiry during science lessons (Hofstein and Lunetta, 2004). One way to overcome this difficulty is to raise students' situational interest. The *Teaching Enquiry with Mysteries Incorporated* (TEMI) project suggested that scientific mysteries are one way to provide such a stimulus. TEMI defined its understanding of mysteries as:

> a phenomenon or event that induces the perception of suspense and wonder in the learner, initiating an emotion-laden 'want to know'-feeling which promotes curiosity and initiates the posing of questions to be answered by enquiry and problem-solving activities. (TEMI Project, 2015, p. 5)

The TEMI project, funded by the EU under the Seventh Framework Programme from 2012 to 2016, suggests that "mysterious" scientific phenomena could have the potential to engage students emotionally in science, by enticing them to solve a given mystery through inquiry (TEMI Project, 2015). The project focuses on four aspects (innovations):

- create curiosity with mysteries
- maintain motivation with showmanship
- inquiry-based science education (IBSE) through the 5E model (engage, explore, explain, expand, evaluate)
- teaching with gradual release of responsibility (GRR).

TEMI suggests that introducing inquiry-based activities focusing on mysteries can be an effective pedagogical tool for promoting situational interest and thus active engagement in inquiry-based activities. The positive influence of using mysterious events in connection with inquiry on students' attitudes toward science has already been reported (Lin, 2014). One such "mystery" used in TEMI involves hydrophobic sand (Peleg *et al.*, 2015).

Whereas building sandcastles with regular sand is well known by students, the hydrophobic sand phenomenon should be unfamiliar to most learners. It thus creates a good and fruitful mystery that, by inducing a cognitive conflict, can generate curiosity and thus situational interest. This phenomenon can be investigated using school-based scientific inquiry in one or two lessons and fits within high-school students' abilities and the related zone of proximal development. The phenomena of hydrogen bonds and hydrophilic and hydrophobic behavior are central to any chemistry curriculum, making the example with hydrophobic sand well justified. The remaining open question is how such a phenomenon can be introduced (Peleg *et al.*, 2015) in the chemistry classroom.

In the TEMI project, it was suggested that one of the main factors that influence students' engagement is the way the teacher presents the mystery (TEMI Project, 2015). There are various ways to present a mystery (such as by showing a video, a demonstration, performing an unexpected experiment, a quiz, *etc.*); TEMI draws on the art of showmanship. Within the context of the TEMI project, showmanship is defined as "the art of making something look interesting and great" (TEMI Project, 2015, p. 36). The TEMI project introduces and conducts inquiry-based learning around mysteries, enriched by using elements of showmanship, such as story-telling, drama, or mime. Because these are uncommon pedagogies for teaching science, showmanship was included as part of the CPD program offered by the TEMI project in the participating countries, among them Israel and Germany.

A theoretical justification for using stories in science classes was pro-posed by Bruner (1985, 1991). He identified two modes of thought and discourse: a paradigmatic, logical mode (a formal method of linking ideas through logic) and a narrative mode (the more familiar form of a story). Formal science naturally relies on the former mode, because it is often seen in scientific papers. Despite the advantages of using stories and narratives, most of the time, science teaching resorts to paradigmatic modes of ex-planation. In general, despite the ubiquitous nature of stories, academic disciplines prefer logical paradigmatic reasoning to narrative reasoning because it is considered to be more "scientific" (Jonassen and Hernandez-Serrano, 2002).

Storytelling can be considered part of drama-based pedagogies (DBP), which include a range of drama-based teaching and learning strategies (Lee *et al.*, 2015). The major features defining DBP are that: (i) it is facilitated by a teacher, a teaching artist, or other facilitator trained in DBP, (ii) it aims at academic and/or psychosocial outcomes for the participating students, (iii) it focuses on process-oriented and reflective experiences, and (iv) it draws on a broad range of applied theatrical strategies.

Lee *et al.* (2015) noted that DBP have a positive and significant impact on achievement outcomes. Despite the overall positive effect of DBP on stu-dents' learning outcomes, little is known about the conditions under which they are effective. This domain is both under-theorized and under-researched (Ødegaard, 2003; Lee *et al.*, 2015). However, there are a few studies reporting the use of stories and DBP in science education. These studies encompass a variety of domains, including electricity (Braund, 1999), molecules and the states of matter (Metcalfe *et al.*, 1984), and mixtures and solutions (Arieli, 2007). In science education, stories are also used to teach argumentation (Erduran and Pabuccu, 2015).

Research findings suggest that drama and storytelling activities might not necessarily improve factual recall (Metcalfe *et al.*, 1984; Ødegaard, 2003), but they can lead to a deeper understanding of the topic being learned (Braund, 1999; Arieli, 2007). Research has also shown that, on the one hand, teachers spontaneously use drama and storytelling activities in the science classroom even if they are untrained in their use and do not consider their pedagogy as DBP (Dorion, 2009). On the other hand, Alrutz (2004) reported that many teachers, while appreciating the potential benefits of DBP, feel a lack of confidence and lack the skills to implement such activities in their classroom.

To effectively implement TEMI, the teachers needed to be adequately prepared. The TEMI CPD was aimed at enabling teachers to introduce mysteries to students by showmanship skills and DBP, such as storytelling. The CPD focused on the five innovations mentioned above, expressed in three types of activities: (i) activities aimed at providing teachers with presentation and storytelling skills, (ii) ready-made inquiry activities and materials, and (iii) construction of classroom activities by the teachers themselves.

A case study in Israel describes a 16-hour CPD program for 14 high-school chemistry teachers (Peleg *et al.*, 2017). The teachers had different levels of experience – from 5 to 15 years of teaching practice. The instructional model of the workshop was that of the GRR approach, in which teachers initially acted as learners and gradually assumed more active roles (Pearson and Gallagher, 1983). The workshop was accompanied by an evaluation study (Peleg *et al.*, 2017). Data were collected by a questionnaire administered online to the teachers 2 to 3 weeks after the last workshop session. The questionnaire included both open and objective-type items intended to capture teachers' reflective experiences regarding the professional development program and how the activities were implemented in class. Two months after the last meeting, attempts were made to schedule semi-structured face-to-face or phone interviews. The final interview sample consisted of 10 teachers.

The interview questions probed how the activities were implemented in class and teachers' reflections regarding the CPD. In addition, seven of the teachers were supervised during their enactment of a mysterious activity in class: five of the teachers implemented activities created by the CPD providers and another two used activities that they had developed themselves. Finally, observation and documentation of discussions, conducted in all of the CPD sessions, were monitored, and written reflective accounts of the activities' implementation were obtained from 10 of the participating teachers.

The evaluation of the workshop showed that stories were a supportive tool that could enrich the pedagogy of inquiry learning. However, there was a mismatch between the teachers' perceptions about themselves as science teachers and their expected image from the students' perspective. Although it can be assumed that most of the teachers read books or watch stories at the movies or on TV, they did not seem to make the connection between the attractiveness of fictitious stories and their potential use in science classes. Some teachers suggested that science teachers have to present a different picture of themselves concerning the validity and authenticity of the information provided by them and that the DBP approach is only worthwhile in, for example, language, music, and art classes.

Science teachers struggle and feel uncomfortable dealing with telling fictitious stories in a classroom involving real science. The teacher may choose to tell students in advance that they will hear a story that is not entirely true, but the students believe that this decreases the story's effect.

The teacher may choose not to announce that the story is fictitious, but many science teachers seem to find this difficult to do. In fact, we want students to "suspend their disbelief", similar to audiences that enter a theater. However, if stories are so effective for memorization, one needs to ask whether the students really want to have a picture of the science teacher only recounting "true" facts and theories. Maybe it is those students who are not intrinsically motivated to study science who need other unconventional approaches, such as mysteries. This may also give science, in general, and chemistry, in particular, a more humanistic flavor that can help overcome the reluctance to study science and diminish the perception of its irrelevance held by many students. However, this will necessitate additional science teacher professional development. Not every science teacher can perform as an actor in class; however, creating and telling a good and motivating story should be possible for everyone. Based on the case study detailed here, there is reason to believe that it is worth enriching science and chemistry education by incorporating elements of DBP (Peleg *et al.*, 2017).

3.5 Summary

- Traditional measures of chemistry teacher professional development are based on top-down dissemination of research, teaching ideas, and materials: media channels, print and online publications, teacher conferences and half- or full-day workshops. If there are no more advanced options for chemistry teachers' professional development, these traditional measures retain their pre-eminence.
- Ideally, top-down professional development should involve long-term strategies to allow teachers to become familiar with new information and approaches and develop ownership. Single-contact events are of limited potential; higher potential can be achieved by repeated face-to-face contact with other teachers who have participated in the same CPD program and with professional development providers.
- Even top-down professional development should take into account the prior knowledge, experiences and beliefs of the participants, make the teachers' prior experience connect to the new content and teaching ideas, and relate the new content or teaching strategies to practical and reflected experience when introducing more advanced teaching strategies. The examples reported in this chapter may serve as patterns for short- and long-term, top-down professional development of chemistry teachers.

References

Abrahams I., (2011), *Practical Work in Secondary Science: A Minds-on Approach*, London: Continuum.

ACS, (1988), *Chemistry in the Community*, Dubuque: Kendall-Hunt.

Top-down Approaches for Chemistry Teachers' In-service Professional Development   57

ACS, (2017), *Chemistry in Context: Applying Chemistry to Society*, 9th edn, New York: McGraw-Hill.

Adams P. E. and Krockover G. H., (1997), Concerns and perceptions of beginning secondary science and mathematics teachers, *Sci. Educ.*, **81**, 29–50.

Adar L., (1970). *Motivation for Learning and Student's Personality*, School of Education, The Hebrew University of Jerusalem, Jerusalem (in Hebrew).

Alrutz M., (2004), Granting science a dramatic license: exploring a 4th grade science classroom and the possibilities for integrating drama, *Teach. Artist J.*, **2**, 31–39.

Arieli B., (2007), The integration of creative drama into science teaching, Ph.D. dissertation, Kansas State University, Manhattan, USA.

Bell B. and Gilbert J., (1994), Teacher development as personal, professional, and social development, *Teaching and Teacher Education*, **10**, 483–497.

Ben-Zvi R., Eylon B. S. and Silberstein J., (1986), Is an atom of copper malleable? *J. Chem. Educ.*, **63**, 64–66.

Ben-Zvi R., Eylon B. S. and Silberstein J., (1987), Students' visualisation of a chemical reaction, *Educ. Chem.*, **24**, 117–119.

Bodner G. M., (2011), Status, contributions, and future directions of discipline-based education research: the development of research in chemical education as a field of study, retrieved from http://www7.nationalacademies.org/bose/DBER_Bodner_October_Paper.pdf.

Bodner G. M., (2014a), Creation of an American Association of Chemistry Teachers, *J. Chem. Educ.*, **91**, 3–5.

Bodner G. M., (2014b), A brief history of chemistry education & ACS's role to support it, *Chem. Solutions*, **1**(1), retrieved from https://teachchemistry.org/periodical.

Bolte C., Streller S. and Hofstein A., (2013), How to motivate students and raise their interest in chemistry education, in Eilks I. and Hofstein A. (ed.), *Teaching chemistry – a studybook*, Rotterdam: Sense, pp. 67–95.

Braund M., (1999), Electric drama to improve understanding in science, *Sch. Sci. Rev.*, **81**(294), 35–41.

Brickhouse N. W. and Bodner G. M., (1992), The beginning science teacher: narratives of convictions and constraints, *J. Res. Sci. Teach.*, **29**, 471–485.

Bruner J., (1985), Narrative and paradigmatic modes of thought, in Eisner E. (ed.), *Learning and Teaching the Ways of Knowing*, Chicago: NSSE, pp. 97–115.

Bruner J., (1991), The narrative construction of reality, *Crit. Inquiry*, **18**(1), 1–21.

Brunkhorst H. K., Yager R. E., Brunkhorst B. J., Apple M. A. and Andrews D. M., (1993), The Salish consortium for the improvement of science teaching preparation and development, *J. Sci. Teach. Educ.*, **4**(2), 51–53.

Buntine M. A., Read J. R., Barrie S. C., Bucat R. B., Crisp G. T., George A. V., Jamie I. M. and Kable S. H., (2007), Advancing Chemistry by Enhancing Learning in the Laboratory (ACELL): a model for providing professional and personal development and facilitating improved student laboratory learning outcomes, *Chem. Educ. Res. Pract.*, **8**, 232–254.

Bybee R. W., (1997), Meeting the challenges of achieving scientific literacy, paper presented at the International Conference on Science Education: Globalization of Science Education, Seoul, Korea.

Constable H. and Long A., (1991), Changing science teaching: lessons from a long-term evaluation of a short in-service course, *Int. J. Sci. Educ.*, **13**, 405–419.

Dillon J. and Osborne J. F., (2007), *Science Education in Europe: Report to the Nuffield Foundation*, London: King's College.

Dorion K. R., (2009), Science through drama: a multiple case exploration of the characteristics of drama activities used in secondary science lessons, *Int. J. Sci. Educ.*, **31**, 2247–2270.

Eilks I., Prins G. T. and Lazarowitz R., (2013), How to organize the classroom in a student-active mode, in Eilks I. and Hofstein A. (ed.), *Teaching Chemistry – A Studybook*, Rotterdam: Sense, pp. 183–212.

Eilks I., Witteck T. and Pietzner V., (2010), Using multimedia learning aids from the Internet for teaching chemistry – Not as easy as it seems? in Rodrigues, S. (ed.), *Multiple Literacy and Science Education: ICTS in Formal and Informal Learning Environments*, Hershey: IGI Global, pp. 49–69.

Erduran S. and Pabuccu A., (2015), Promoting argumentation in the context of chemistry stories, in Eilks I. and Hofstein A. (ed.), *Relevant Chemistry Education*, Rotterdam: Sense, pp. 143–161.

European Commission, (2004), *Scienter - Identification and dissemination within Europe of best practices in the context of science teaching that places science and technology into meaningful contexts (executive summary)*, Brussels: EU.

European Commission, (2007), *Science Education Now: A Renewed Pedagogy for the Future of Europe*, Brussels: EU.

Gabel D. L. and Sherwood R. D., (1981), Using facilitating problem solving in high school chemistry, paper presented at the NARST conference, New York, USA.

Guskey T. R., (2003), What makes professional development effective? *Phi Delta Kappa*, **84**, 748–750.

Hall G. and Loucks S., (1981), Program definition and adaptation implications for inservice, *J. Res. Dev. Educ.*, **14**(2), 46–58.

Herron J. D., (1975), Piaget for chemists, *J. Chem. Educ.*, **52**, 146–150.

Hofstein A., Eilks I. and Bybee R., (2011), Societal issues and their importance for contemporary science education: a pedagogical justification and the state of the art in Israel, Germany and the USA, *Int. J. Sci. Math. Educ.*, **9**, 1459–1483.

Hofstein A. and Kempa R. F., (1985), Motivating strategies in science education: attempt at an analysis, *J. Chem. Educ.*, **7**, 221–229.

Hofstein A. and Lunetta V. N., (2004), The laboratory in science education: foundations for the twenty-first century, *Sci. Educ.*, **88**, 28–54.

Hutchinson K., Bodner G. M. and Bryan L., (2011), Middle-and high-school students' interest in nanoscale science and engineering topics and phenomena, *J. Pre-College Eng. Educ. Res.*, **1**(1), 30–39.

Jonassen D. H. and Hernandez-Serrano J., (2002), Case-based reasoning and instructional design: using stories to support problem solving, *Educ. Techn. Res. Dev.*, **50**(2), 65–77.

Kempa R. F., (1983), Developing new perspectives in chemical education, paper presented at the 7th International Conference in Chemistry, Education, and Society, Montpellier, France.

Kempa R. F. and Diaz M. M., (1990), Students' motivational traits and preferences for different instructional modes in science education: part 2, *Int. J. Sci. Educ.*, **12**, 205–216.

Kind V., (2009), Pedagogical content knowledge in science education: perspectives and potential for progress, *Stud. Sci. Educ.*, **45**, 169–204.

Kolesnick W. B., (1978), *Motivation, Understanding and Influencing Human Behavior*, Boston: Allyn and Bacon.

Krajcik J., Mamlok R. and Hug B., (2001), Modern content and the enterprise of science: science education in the 20th century, in Corno L. (ed.), *Education Across a Century: The Centennial Volume*, Chicago: NSSE, pp. 205–238.

Lee B. K., Patall E. A., Cawthon S. W. and Steingut R. R., (2015), The effect of drama-based pedagogy on PreK-16 outcomes: a meta-analysis of research from 1985 to 2012, *Rev. Educ. Res.*, **85**(1), 3–49.

Lin J.-L., (2014), Learning activities that combine science magic activities with the 5E instructional model to influence secondary-school students' attitudes to science, *EurAsia J. Math. Sci. Techn. Educ.*, **10**, 415–426.

Lortie D., (1975), *School Teacher: A Sociological Study*, Chicago: University of Chicago.

Loucks-Horsley S. and Matsumoto C., (1999), Research on professional development for teachers of mathematics and science: the state of the scene, *Sch. Sci. Math.*, **99**, 258–271.

Loucks-Horsley S., Stiles K. E., Mundry S. and Hewson P. W., (2009), *Designing Professional Development for Teachers of Science and Mathematics*, Thousand Oaks: Corwin.

Lunetta V. N., Hofstein A. and Clough M. P., (2007), Learning and teaching in the school science laboratory: an analysis of research, theory, and practice, in Abell S. K. and Lederman N. G. (ed.), *Handbook of Research on Science Education*, Mahwah: Lawrence Erlbaum, pp. 393–441.

Mamlok-Naaman R., Blonder R. and Hofstein A., (2010), Providing chemistry teachers with opportunities to enhance their knowledge in contemporary scientific areas: a three-stage model, *Chem. Educ. Res. Pract.*, **11**, 241–252.

Mamlok-Naaman R., Hofstein A. and Penick J., (2007), Involving teachers in the STS curricular process: a long-term intensive support framework for science teachers, *J. Sci. Teach. Educ.*, **18**, 497–524.

Metcalfe R. J. A., Abbott S., Bray P., Exley J. and Wisnia D., (1984), Teaching science through drama: an empirical investigation, *Res. Sci. Techn. Educ.*, **2**, 77–81.

Munby H., Russell T. and Martin A. K., (2001), Teachers' knowledge and how it develops, in Richardson V. (ed.), *Handbook of Research on Teaching*, Washington: AERA, pp. 877–904.

National Research Council, (2013), *Next Generation Science Standards: For States, by States*, Washington: National Academies Press.

Novak J. D., Ring D. J. and Tamir P., (1971), Interpretation of research findings in terms of Ausubel's theory and implications for science education, *Sci. Educ.*, **55**, 483–519.

Ødegaard M., (2003), Dramatic science. A critical review of drama in science education, *Stud. Sci. Educ.*, **39**, 75–101.

Orgill M. and Bodner G. M., (2014), ACELL project: advancing chemistry by enhancing learning in the laboratory, in *Abstracts of papers of the American Chemical Society*, vol. 248.

Osborne J. F. and Dillon J., (2008), *Science Education in Europe: Critical Reflections*, London: Nuffield.

Pearson P. D. and Gallagher M. C., (1983), The instruction of reading comprehension, *Contemp. Educ. Psych.*, **8**, 317–344.

Peleg R., Katchevich D., Yayon M., Mamlok-Naaman R., Dittmar J. and Eilks I., (2015), The magic sand mystery, *Sci. School*, **32**, 37–40.

Peleg R., Yayon M., Katchevich D., Mamlok-Naaman R., Fortus D., Eilks I. and Hofstein A., (2017), Teachers' views on implementing storytelling as a way to motivate inquiry learning in high-school chemistry teaching, *Chem. Educ. Res. Pract.*, **18**, 304–309.

Schleicher A., (2009), Seeing learning outcomes in Israel through the prism of global comparisons, retrieved from http://cms.education.gov.il/EducationCMS/Units/Rama/MaagareyYeda/MaagareiYeda_Mazagot_heb.htm?WBCMODE=presentationunpublishedmavo.htm.

Sharan S. and Hertz-Lazarowitz R., (1980), A group investigation method of cooperative learning in the classroom, in Sharan S. (ed.), *Cooperation in Education*, Provo: Brigham Young University Press, pp. 14–46.

Shulman L. S., (1986), Those who understand: knowledge growth in teaching, *Educ. Res.*, **15**(2), 4–14.

Simmons P. E., Emory A., Carter T., Coker T., Finnegan B., Crockett D., Richardson L., Yager R., Craven J., Tillotson J., Brunkhorst H., Twiest M., Hossain K., Gallager J., Duggan-Haas D., Parker J., Cajas F., Alshannag Q., McGlamery S., Krockover G., Adams P., Spector B., La Porta T., James B., Rearden K. and Labuda K., (1999), Beginning teachers: beliefs and classroom actions, *J. Res. Sci. Teach.*, **36**, 930–954.

Stuckey M., Hofstein A., Mamlok-Naaman R. and Eilks I., (2013), The meaning of 'relevance' in science education and its implications for the science curriculum, *Stud. Sci. Educ.*, **49**, 1–34.

Taitelbaum D., Mamlok-Naaman R., Carmeli M. and Hofstein A., (2008), Evidence-based continuous professional development (CPD) in the inquiry chemistry laboratory (ICL), *Int. J. Sci. Educ.*, **30**, 593–617.

TEMI Project, (2015), *How Using Mysteries Supports Science Learning*, London: TEMI.

Tuvi-Arad I. and Blonder R., (2010), Continuous symmetry and chemistry teachers: learning advanced chemistry content through novel visualization tools, *Chem. Educ. Res. Pract.*, **11**, 296–301.

US Department of Education, (2016), High school coursetaking, in *The Condition of Education 2016*, Washington: National Center for Education Statistics, 2016-144.

Wade R. K., (1985), What makes a difference in inservice teacher education? A meta-analysis of research, *Educ. Leadership*, **42**(4), 48–54.

Wellington J., (1991), Newspaper science, school science: friends or enemies?, *Int. J. Sci. Educ.*, **13**, 363–372.

Wischow E. D., Bryan L. and Bodner G. M., (2013), Secondary science teachers' development of pedagogical content knowledge as result of integrating nanoscience content in their curriculum, *COSMOS*, **8**, 187–209.

Wood D., (2007), Teachers' learning communities: catalyst for change or a new infrastructure for the status quo, *Teach. Coll. Rec.*, **109**, 699–739.

Yager R. E. and Simmons P., (2013), Results of the Salish projects: summary and implications for science teacher education, *Int. J. Educ. Math. Sci. Techn.*, **1**, 259–269.

CHAPTER 4

Cases of Bottom-up Professional Development for Chemistry Teachers

Regarding in-service teacher education, this chapter focuses on "bottom-up" (as opposed to "top-down") models for the professional development of chemistry teachers. The chapter discusses three professional development cases from Israel that involved chemistry teachers as curriculum developers, classroom innovation by teachers while cooperating in an in-service workshop on their classroom activities, and teachers' professional development in learning communities.

4.1 Teachers' Involvement in Curriculum Development and Implementation

In their book *Designing Professional Development for Science and Mathematics Teachers*, Loucks-Horsley *et al.* (2010) listed 15 specific and effective professional development strategies. It is beyond the scope of this chapter to list them all, but it is worth mentioning those in which the key characteristics are intensive involvement of science teachers in changing the curriculum, using focus groups, study groups, action research, curriculum development, and curriculum implementation.

Loucks-Horsley *et al.* (1998) referred to these strategies (p. 43). Regarding curriculum development and adaptation, they wrote: "creating new instructional materials and teaching strategies or tailoring existing ones to meet the learning needs of students". Regarding curriculum implementation, they wrote: "learning, using, and refining the use of a particular set of

Advances in Chemistry Education Series No. 1
Professional Development of Chemistry Teachers: Theory and Practice
By Rachel Mamlok-Naaman, Ingo Eilks, George Bodner and Avi Hofstein
Published by the Royal Society of Chemistry, www.rsc.org

instructional materials in the classroom". Clearly, these two professional development strategies, namely, the development of an innovation and implementing it in the chemistry classroom, complement each other. Both strategies have potential for the intensive involvement of chemistry teachers as part of their professional development and thus, they should be conducted in parallel.

The goal of curriculum development as a professional development strategy is to have teachers create new learning materials or instructional techniques that will be implemented (also as a professional development strategy) in the science (chemistry) classroom. The teachers' curriculum *development* and *implementation* of professional development strategies involve the following aspects: they learn new science content (content knowledge; CK) and aligned pedagogies (pedagogical content knowledge; PCK); they collaborate with peers, experts, and professional development providers; they plan assessment strategies aligned with the content and pedagogy (see Chapters 2 and 3 of this book). In addition, during the professional development process, while implementing the learning materials, the teachers are provided with opportunities to reflect on their classroom experiences. Loucks-Horsley *et al.* (1998) suggested that these activities have the potential to enhance teachers' professional growth, eventually leading to more effective classroom teaching and learning practices. Encouraging the teachers to reflect on their experiences during the professional development process and the implementation of the modules, and exploring those reflections, may serve as a tool for describing the process of gaining a sense of ownership.

Regarding teachers' reflections, teachers need to familiarize themselves with new ideas and also understand the implications, both for themselves as teachers and for their learners in the classroom, before they adopt and adapt them. If the new approach differs greatly from their previous practice, they need to be involved in reshaping their own beliefs regarding science teaching and learning. This involves both considering core principles and issues and contextualizing them in developing effective pedagogical approaches that can be put into practice.

One common theme underlying recent reports on science education in general, and chemistry education in particular, is that the content of traditional school science and its related pedagogical approaches are not aligned with the interests and needs of either society or most students. Many students do not find their science classes interesting, motivating, or relevant (Stuckey *et al.*, 2013). These claims are especially valid regarding those students who, in the future, will not embark on a career in science or engineering but will need a basic understanding of science and technology at a personal and functional level as literate citizens (Hofstein *et al.*, 2011).

One key problem seems to be that globally, few science programs teach how science is linked to those issues that are relevant to students' lives, the environment, or their role as future citizens. As a result, many students are unable to participate in societal discussions about science and its related technological applications (Hofstein *et al.*, 2011). In this chapter, we attempt to highlight various issues related to chemistry teachers' development of a sense of ownership through long-term professional development for new chemistry content and pedagogical approaches. We suggest reflective methods to identify and reveal teachers' development of a sense of ownership.

4.2   Chemistry Teachers as Curriculum Developers

It is reasonable to assume that involving teachers in the process of curriculum development and its related implementation in the school system will reduce the mismatch, or gap, between the intent of curriculum innovation and the teachers' needs and concerns. Learning materials that result from teachers' intensive involvement have more potential to be adopted effectively in today's schools. In general, the involvement of leading teachers in the long-term professional development and implementation of new curricula leads to effective customizations aligned with the developers' original rationale, while still allowing teachers to respond to local needs, and the unique character of the school and its related classroom learning environments (Loucks-Horsley *et al.*, 1998). In addition, encouraging teachers to reflect on their experiences during their professional development and implementation of the new lesson plans, and exploring those reflections, may serve as a tool for teachers' development of a sense of ownership.

The reflective teaching practice is a process in which teachers think about their teaching and analyze how something was taught and how the practice might be improved or changed for better learning outcomes. Some points of consideration in the reflection process might include thinking about what is currently being done, why it is being done, and how effectively students are learning (Obaya, 2003). Teachers need to familiarize themselves with new ideas, and understand the implications for themselves as teachers and the benefits for their students before they adopt and adapt them. If the new approach differs greatly from their previous practice, the teachers need to be involved in reshaping their own beliefs regarding science teaching and learning. This involves considering and contextualizing core principles and issues to develop effective pedagogical approaches in theory and in practice.

The following section provides some insights into the idea of a workshop focusing on "teachers as curriculum developers" (Mamlok-Naaman *et al.*, 2007). The participants were 10 teachers from 10 schools located in central Israel who had at least 10 years of high-school science teaching experience, mainly in grades 10–12. Their scientific backgrounds differed, and included chemistry, biology, agriculture, nutrition, technology, and physics. The teachers participated in the development of alternative assessment tools in the context of implementing a new science curriculum for senior high-school students, namely "Science for All", a science–technology–society (STS)-type curriculum (see Chapter 6 of this book). The "Science for All" program was developed as part of a more comprehensive reform in science education that has been evolving in Israel since 1992. In the early 1990s, the Israeli Ministry of Education set up a committee that considered the need to make science an integral part of the education of all citizens (Tomorrow 98, 1992). In 1992, the recommendations of the committee were accepted, and the government committed to the idea that science would be taught to all of the country's high-school students. It was also decided that different programs would be taught in grades 10–12 to science majors, and to non-science-oriented students who did not choose to major in any of the science disciplines (biology, chemistry, or physics).

The "Science for All" program consisted of a set of 15 modules (35–40 hours each) that all had a STS structure and content. Each module focused on a specific scientific topic, *e.g.*, *Energy and the Human Being* (Ben-Zvi, 1998, 1999), *Science: An Ever-Developing Entity* (Mamlok, 1998), and *Brain, Medicine, and Drugs* (Cohen *et al.*, 2004). The *Energy and the Human Being* module tried to clarify some issues concerning beliefs and mis-conceptions about energy. *Science: An Ever-Developing Entity* was designed to develop an understanding of the nature of science by using historical ex-amples (see Chapter 2 of this book). In this way, science was presented as a continuously developing enterprise of the human mind in the context of the historical development of our understanding of science (Mamlok *et al.*, 2000). *Brain, Medicine, and Drugs* focused on several selected aspects of brain research, and their relationship to human behavior and emotions.

The workshop participants met eight times for 4 hours every other week. Two science education researchers conducted the workshop and the re-search associated with it. The workshop was initiated to address the tea-chers' questions: "What strategies should we use to teach STS modules, and how should we assess the students who are studying such modules?" When teaching the modules, the teachers were expected to use a wide range of pedagogical interventions and instructional techniques to cope with a wide range of student abilities, interests, and means of motivation. Moreover, the implementation of a STS program with a wide spectrum of learning goals necessitated matching instructional techniques and assessment tools to measure students' achievements in, and progress toward each learning goal (Hofstein *et al.*, 2006).

Each of the workshop participants was asked to teach the "Science for All" program in one class. The teachers had already taught the previously de-scribed "Science for All" modules, but had difficulty using a variety of teaching strategies in general, and in grading and assessing their students in particular. As already noted, the workshop was initiated to address the needs of the teachers who implemented the "Science and Technology" program with respect to their teaching strategies and related assessment methods. Therefore, the workshops focused on guiding the participating teachers in using a variety of teaching strategies, and in the development of auxiliary assignments for their students, together with assessment tools. The as-sessment tools used in this workshop consisted of detailed checklists (rubrics) and rating scales. In the first three meetings, the participating teachers were exposed to lectures and to activities related to the alternative assessment tools and methods, and especially in the way in which they should become used to working with rubrics. Each teacher prepared the assignments for his or her students, as well as the appropriate assessment tools, which included tests, quizzes, and assessment guides for carrying out mini-projects, writing essays, and critical reading of scientific articles. All of the assignments were developed in stages, each of which required consideration and an analysis of assessment criteria as well as scoring. These assignments were administered stage by stage at school. The students were involved in the assessment methods and their relative weights associated with each assessment. This ongoing assessment process provided them with more control over their achievements, because they were aware of the assessment method, the weight percentage for each of the assessment components, and the final grade. At each stage, the students submitted their papers to the teacher for comments, clarification, and assessment. The students met with the teachers before and after school for extra instruction and consultation. The students reflected on their work and ideas at each stage, and followed their teachers' comments on a detailed checklist, correcting them accordingly.

Samples of the students' assignments were brought to the workshop for further analysis, involving both the coordinators and their colleagues. The group discussed the revision of the rubrics, and agreed on the percentage (weight) allocated to each of the assignment components. They also agreed on the criteria for performance levels, so as to grade the students as objectively as possible. The different components of the workshop are presented in Figure 4.1, specifically: (i) discussions of the teaching methods, (ii) preparation of learning and auxiliary materials and assessment tools, (iii) development of rubrics/criteria for assignment assessment, (iv) analysis of samples of students' assignments, and (v) improvement and revision of the rubrics according to the samples of the students' assignments.

Each teacher developed a variety of student assignments and assessment tools. Assessment criteria for the assignments were suggested and discussed in the workshop in terms of both their content and weight. The following two assignments serve as examples.

4.2.1 Critical Reading of Scientific Articles Published in Newspapers or Other Media and Original Scientific Articles

Scientific articles published in daily newspapers and scientific newspapers can serve as an important source for enrichment and for making the studied subject more authentic and up to date (Wellington, 1991). These articles are originally written by scientists; more specifically, they consist of scientists' reports on their research work, which are ultimately published in

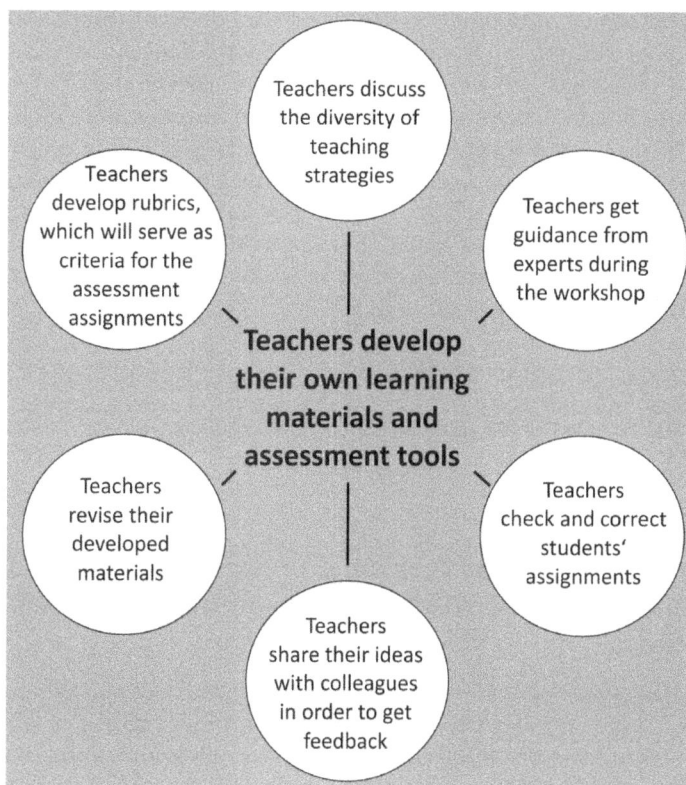

Figure 4.1 The different stages of the workshop (Mamlok-Naaman *et al.*, 2007).

professional journals (Yarden *et al.*, 2001). To use them in high school, however, they need to be modified into a popular, easily readable version. Note that critical reading of daily as well as scientific newspaper articles is thought to contribute to developing a student who is literate in the sciences (Norris and Phillips, 2003). Each student in the class had to choose an article from a collection provided by the teacher. The students were also provided with a written guide for critically reading the paper (Levy Nahum *et al.*, 2004). The articles given to the students dealt with the following topics: Important Elements, The Discovery of the Rare Earth Elements, Chemistry in the Bible, Thermodynamics and Spontaneity, The Story of Energy, and Chemical Aspects of Atmospheric Pollution.

The articles underwent a simplification stage to adapt them to the students' reading ability and to their chemistry background. For the purpose of simplification, the article was organized (and written) in sections that followed the organization of an authentic research paper, namely abstract, introduction, research methods, results, and summary. The introduction presented the necessary scientific background. The introduction also provided the students with a glossary of new and unfamiliar words, equipment, *etc.*, such as semiconductors and resistors. The research methods section introduced the students to the methods used by the scientists in their work. At the end of the article, we wrote a short summary containing the main ideas incorporated in the article. The results were presented on a graph that showed the different experimental conditions. Article selection was based on the assumption that they presented topics that could be characterized in terms of "frontiers in chemistry", were relevant, and had technological applications; these were also topics that the workshop participants thought would be of interest to the students. The students were asked to read the article and then to:

- identify at least five scientific concepts whose meaning was unknown to them
- compile questions that criticized the article's contents
- answer the compiled questions.

4.2.2   Writing an Essay Focusing on Scientists and Their Discoveries, Entitled "The Person behind the Scientific Endeavor"

To help students write an essay on "The Person behind the Scientific Endeavor", the teachers introduced them to the biographies of numerous eminent scientists from different periods. These individuals developed scientific theories that often contradicted those that had been previously accepted. The students were asked to describe in detail the lives of these scientists and the discoveries they made. They also created materials that characterized "their" scientists: a picture of the scientist accompanying an essay that the students had written. The students used internet resources, and the teachers helped them with references dealing with the history of science (Conant and Nash, 1964; Priesner, 1991; Seybold, 1994; Rayner-Canham and Rayner-Canham, 1998). The class then constructed a display along a timeline to place events, scientists, and theories in their appropriate historical perspective. Thus, the students all felt that each scientist had been given an honorable place in the history of science.

An evaluation study was conducted during the workshop and at its completion. The main goal of the study was to evaluate the outcomes of the workshop and to determine whether its objectives had been attained. The research tools consisted of: (i) an attitude questionnaire administered to participating teachers, (ii) structured interviews with the teachers, (iii) structured interviews with the students, and (iv) an attitude questionnaire administered to the students. One of the authors of this study wrote a protocol of the discussions held during the meetings. From reading the minutes of the meetings, the researchers could learn about the teachers' conceptions regarding student learning and their learning environment, their attitudes toward a variety of teaching and assessment strategies, as well as their specific difficulties. The minutes helped clarify the data collected from the interviews and were analyzed according to issues that were revealed during the workshop meetings. The researchers also visited the schools at the completion of the STS unit and interviewed a sample of students from the 10 teachers who had participated in the workshop. In each class, the researchers interviewed four students who were chosen by their teachers according to their achievements to include two high achievers and two low achievers. A total of 40 students were therefore included in these interviews, representing 13% of all students.

Additional relevant data referred to the main goal of the study, namely, determining whether the objectives of the workshop had been attained. We used teacher self-report questionnaires and teacher interviews based on the literature (Lawrenz, 2001), which claims that such instruments can be regarded as valid and reliable if they are administered and the data collected at times when a person's almost immediate response can be obtained. Interviews with the students focused on the affective, rather than cognitive, aspects of learning.

Based on the results of the questionnaires and the interviews, we concluded that the teachers who participated in the workshop had gained confidence in using the teaching strategies and assessment methods of this new interdisciplinary curriculum. The interviews with students revealed that their active involvement in their own assessment improved their sense of responsibility for their achievements. The variety of assignments enabled them to be better at certain assignments and less successful with others. In conclusion, it was found that introducing a new interdisciplinary curriculum is facilitated by professional development programs that stimulate teachers' creativity and diversify the instructional strategies used in the classroom. Such skills should improve their ability to understand the goals, strategies, and rationale of the curriculum, as well as their students' learning difficulties.

This workshop was initiated to help a group of teachers who had asked for support in implementing a new science curriculum in both teaching and assessment strategies. It brought together teachers from different backgrounds (biology, chemistry, physics, and agriculture) with one common objective, thus enabling them to contribute to and enrich one another. Two main themes emerged from the participants' responses in the questionnaires and in the interviews, and from the minutes of the meetings: the teachers became more self-confident in the teaching and assessment methods of this new interdisciplinary curriculum and were motivated to try new content and teaching strategies. Moreover, they better understood the advantages of the alternative assessment methods and were better prepared to use them. We believe that the teachers who were involved in this process were satisfied with, and felt pride in, their work and their accomplishments.

The new curriculum materials also appeared to be effective vehicles for teacher learning (Bybee and Loucks-Horsley, 2000). The teachers were involved in the development of the learning materials as well as the teaching strategies and assessment tools, which had to be adequately tailored to the students' cognitive and affective characteristics (Ben-Peretz, 1990). Hopefully, in the future, these teachers will serve as leaders and coordinators for similar workshops, and support those who will teach the STS modules and use the alternative assessment methods. As a result, the interest of these teachers' students in the process of learning increased, as well as their satisfaction from the learning materials, learning strategies, assessment methods, and ongoing dialog with their teachers. As already noted, because the students who studied the STS program were not science-oriented and their interest in scientific topics was limited, the variety of assignments enabled them to succeed with certain assignments and to do less with other activities. These results are in alignment with the main goal of the reform in science education in Israel – the need to make science an integral part of the education of all citizens (Tomorrow 98, 1992):

> Modern socioeconomic problems require an understanding of their scientific background. Other questions arise when we discuss the division of resources and world wealth, different environmental issues and other topics that require the individual to demonstrate an understanding based on having acquired a basic education in the sciences. (p. 3)

The teachers who participated in the workshop became aware of the difficulties that could arise regarding the validity and reliability of the assessment tools. Thus, they made great efforts to improve and revise the assignments and rubrics based on the students' completed assignments. In fact, their anxiety about the alternative assessment methods gradually diminished when they realized that continuous assessment of students' progress and achievements, consisting of detailed and clear assessment instructions, could present a broad, valid, and reliable picture of their students' knowledge and abilities.

To attain a wide range of assessment models, sufficient time is needed to construct a supporting framework for science teachers (Westerlund *et al.*, 2002). Indeed, the teachers in the workshop were continuously supported and assisted by the workshop coordinators.

4.3   Classroom Innovation by Teachers While Cooperating in In-service Workshops on Their Classroom Activities Within the PROFILES Project

The goal of the PROFILES project (Professional Reflection Oriented Focus on Inquiry-based Learning and Education through Science; FP7 project on Science in Society), involving 20 countries in Europe, was to make science learning more relevant to the learner by promoting more meaningful science education. To achieve this goal, the PROFILES project focused on four key aspects designed to vary the classroom learning environment by using different pedagogical approaches and instructional techniques (Bolte *et al.*, 2012):

- Making chemistry learning more relevant through means of meeting students' interests (for a detailed discussion about the meaning and scope of the term "relevance", see Stuckey *et al.*, 2013).
- Using a context-based chemistry education approach (Hofstein and Kesner, 2006).
- Educating learners through ideas related to scientific literacy for all students (Holbrook, 2005).
- Using a student-centered, inquiry-based science education (IBSE) approach.

One of the key features of the PROFILES project was the recognition that the novel approach being advocated differed significantly from traditional chemistry teachers' professional experiences in most of the participating countries. Teachers were asked to change their beliefs regarding science teaching and learning, which involved reconsidering core principles and issues as well as contextualizing them in developing practices and approaches.

The challenge of the innovations proposed in the PROFILES project consisted of developing activities that would be needed to support the teachers when they returned to their schools. Professional development sessions were not going to be enough.

The supportive climate of the teachers' meetings was believed to be an integral component of the process of getting changes integrated into routine chemistry teaching practice. Concern was raised during the planning stage that the intervention aspect of the professional development was being put into action before the goals of the PROFILES project could be reflected upon to serve as a further key component of the professional development program. The key goal was to enhance science teachers' skills related to IBSE, and decision making related to societally relevant science learning (Bolte *et al.*, 2012).

Teachers involved in the PROFILES professional development project in Israel were supported by a carefully planned 1-year-long workshop structure. The professional development consisted of four partially overlapping steps: the teacher as learner, the teacher as teacher, the teacher as reflective practitioner and finally, for some of the teachers who were responsible for developing a high-level sense of ownership, the teacher as leader.

The first two steps were targeted throughout the professional development programs for most of the teachers. The main goal of these features was to support teachers in promoting students' learning processes through inquiry. The first component was concerned with enhancing the teachers' CK. The second component (the teacher as teacher) involved the teachers in developing instructional techniques, pedagogical interventions, and assessment methods aligned with the content. The third component, namely, the teacher as reflective practitioner, was fundamentally based on the teachers' classroom experiences in using the altered approaches in practice.

The professional development initiative in PROFILES provided the teachers with opportunities to reflect on their classroom experiences and included pedagogical issues, accomplishments, concerns, and difficulties. In this capacity, the participants and professional development providers used a dynamic platform for feedback and reflection. Their peers and professional development providers suggested ways to improve classroom performance, time management, and the classroom (and laboratory) learning environment. The fourth feature involved developing a high-level sense of ownership regarding the philosophy and pedagogical approaches of the PROFILES project because the development of ownership is one of the most important variables related to developing leadership among participating teachers. Fullan (1991) defined teacher leadership as the teachers' ability to bring about a change in the educational system. The PROFILES project assumed that those chemistry teachers who developed a high level of ownership had the potential to disseminate ideas about PROFILES, be involved in subsequent in-service professional development initiatives, and present the ideas of PROFILES to other teachers in their school or in the region.

Two cycles of professional development, in two consecutive years, were conducted as part of an in-service training program, each including 25 upper secondary school chemistry teachers. The teachers came from urban schools with an average of at least 8 years' experience in teaching chemistry in the upper levels of secondary school (10th to 12th grade). Most of the teachers had been involved in short-term in-service professional development consisting of 4–5 days of teacher training aimed at familiarizing the chemistry teachers with CK and PCK in using a new approach for teaching chemistry (Barnea *et al.*, 2010).

In total, the duration of each of the professional development cycles was approximately 84 hours. Twenty-five per cent of the professional development was conducted during the summer preceding the school year. The remainder of the program was conducted throughout the 2012 and 2013 academic years, either face-to-face or as online activities. During each of the academic years, the emphasis was on activities related to the chemistry teachers' professional development. More specifically, the teachers:

- learned about the rationale underlying PROFILES, the importance of IBSE, the meaning and interpretations of relevance, and modules developed in previous projects
- studied issues related to high-order learning skills in depth, including IBSE, decision making, and other key pedagogical and instructional issues and topics
- created a scientific background related to the module that they had selected for further in-depth investigation and development.

The intensive 3-day summer course served primarily as a vehicle for motivating the teachers to commit themselves to the PROFILES program. Most teachers felt the need to build on their learning through professional development, to develop new chemistry-based modules and not just use those that were recommended by the project on the internet (*e.g.*, the PARSEL project). Indeed, the professional development providers were encouraged by this attitude because it clearly indicated the teachers' involvement in the project, which was aligned with the professional development model: "*the teacher as curriculum developer*", a "bottom-up" approach (Loucks-Horsley *et al.*, 1998). This was assumed to support the potential of enhancing teachers' self-efficacy and later served as an approach to gain evidence for teacher ownership of the ideas from PROFILES. The following is a quote from one of the participants, in writing about his personal reflections on the development and implementation of the "sunscreen protection" module):

> I felt right from the beginning that our group is going to attain the philosophy of PROFILES. In the introductory course during the summer, we had an opportunity to investigate the scenario of the module. This was done in an unusual and very interesting way. We had to be creative in preparing the poster [for the initial presentation]. The preparation involved all of the group members and included both hands-on as well as minds-on activities.

During the professional development, about 45% of the meeting time was conducted online and the other 55% was conducted face-to-face. Teachers were actively involved in developing the learning materials. They also implemented these materials in their schools, which gave them an opportunity to reflect on their experiences. As already noted, the professional development program served as a platform for feedback from peers and the professional development providers. The development of the modules involved several stages based on collaborative efforts with peers and the professional development providers:

- choosing the theme of the module in alignment with the chemistry curriculum and the students' abilities and interests
- collecting correct and valid scientific backgrounds
- designing pedagogical interventions aligned with content and context.

All of these stages represented the key issues and pedagogies of PROFILES, namely, meeting interest, inquiry interventions, and decision making. There were important benchmarks in developing the modules: the design of the first scenario and the design of the inquiry-based activity, which should lead to the decision-making process, and maintaining the timing that was given by the professional development providers.

The developed modules were all original; they were not adapted from other projects. Among the modules that were developed and implemented in the chemistry classrooms were:

- Biodiesel: can used oil be the next-generation fuel?
- Hazard from the sun: which sunscreen should we choose?
- "To drink or not to drink": the issue of drinking alcoholic beverages
- Energy drinks: do they give you opportunities to fly?
- Plastic: reduce the use!

Throughout the professional development program, several methods were used to determine the teachers' development of a sense of ownership related to the PROFILES project in which they were involved. Data were collected using three different tools: (i) an open-ended question, providing teachers with an opportunity to reflect and (ii) reflective essays written by the teachers during the workshop.

The open-ended question asked the teachers to complete the sentence: *"I feel ownership toward PROFILES because _____."* Teachers' claims were sorted and analyzed into six categories: (i) empathy with the project's rationale, (ii) promotion of the teacher's image among peers, (iii) promotion of the teacher's personal image in his/her science class, (iv) willingness to continue working on the project, (v) sharing and disseminating ideas among peers, and (vi) teacher's professionalism. These categories were also used to analyze the data collected by the second instrument, the reflective essays written by the teachers during the workshop: teachers were asked to reflect on their work, using essays, at different times during the workshop.

The following quotes represent some of the data collected from the reflective essays and our interpretation of the different categories regarding a sense of ownership.

- Category 1. Empathy with the project's rationale:
  I believe that teaching chemistry using IBSE as a central pedagogical approach is the best approach for teaching my chemistry students. The combination of teaching relevant ideas (topics) using inquiry and involving the students in the process of decision-making is very motivating and eventually will encourage the students to enroll in more advanced chemistry studies.
- Category 2. Promotion of the teacher's image among peers:
  The coordinator no longer thinks that it is unnecessary to implement the modules; she wants to do the activity in the gifted students class.
- Category 3. Promotion of the teacher's personal image in his/her science class:
  I feel that the students appreciate me more than before. Many times I heard my students talking about me and about the module to their friends. . .I was excited.
- Category 4. Willingness to continue working on the project:
  When we heard that another professional development cycle was being initiated, I decided to go for another round to have the opportunity to

be involved in the development and implementation of another module.
- Category 5. Sharing and disseminating ideas among peers:
  The next assignment in which I want to be involved is to disseminate the module (that we developed in the professional development workshop) among other teachers in other schools.
- Category 6. Teacher's professionalism:
  In addition to enriching me in teaching, educational issues, and a new pedagogy for teaching, developing new units added a challenge, interest, as well as intellectual and personal development.

Analysis of the teachers' individual reflective essays showed that most of them:

- reported a sense of satisfaction regarding the use of the modules that they had developed personally
- identified with the rationale of the project (development and implementation)
- mentioned that they were involved in the dissemination of PROFILES outside their own school.

Many teachers suggested that they:

- involved the principal of their school in the PROFILES project
- published articles for teachers about their experience in the design and implementation of their modules.

Teachers were involved in these activities because they were most suitable for aligning the content and pedagogy for their schools and students. Developing the teachers' sense of ownership was one of the main targets for developing leadership among the teachers. It was postulated that, in the future, these teachers would be able to actively support the development and implementation of reforms in how chemistry is taught and learned by their peers. This study showed that most of the teachers who attended the PROFILES workshop developed a high sense of ownership toward the project and its related pedagogies.

After participating in the PROFILES workshop, the teachers claimed that although the workshop was demanding, the timeline and goals were clear, and they felt that the atmosphere was supportive and open, and considered their needs. Most of them were willing to continue and disseminate the project. Most of the teachers claimed that the PROFILES workshops that they had attended could serve as a model for further professional development aimed at improving science teaching and learning.

We tried to determine the degree of self-efficacy gained by the teachers as a result of the professional development program. Two components were essential for this: the "teacher as teacher" using the ideas from PROFILES and the "teacher as reflective practitioner" who implements the ideas from PROFILES. The first involved the teacher in the project and the second aided teachers in understanding the importance of their involvement. One teacher answered the question: "What I enjoyed most was implementing the bio-diesel module as follows:

> This is mine, I am part of it. I'm excited about the start-up and I had butterflies in my stomach...because after working so much here I reached the final stage where I could implement my module – this is the most important part – to see whether the students would enjoy the activity, whether it would go smoothly; would they feel excited that the module had been developed for them?

We believe that some of the activities that we initiated catalyzed further developments through which teachers provided evidence of real ownership of the PROFILES philosophy and approach:

- Encouraging teachers to present their modules at the national conference of chemistry teachers and sharing their reflections with their peers following the presentation.
- Encouraging teachers to write articles for the journal for chemistry teachers in Israel.
- Encouraging teachers to become partners in writing the newsletter, and referring to all the modules in the newsletter, including pictures of the teachers.
- Providing a forum for teachers to share their modules and get feedback from their peers.

4.4   Teachers' Professional Learning Communities (PLCs)

Another approach for promoting bottom-up professional development for chemistry teachers involved the creation of teachers' PLCs. These are groups of professionals who jointly examine and discuss their knowledge and practice with the aim of improving professionally.

Theoretically, the models of PLCs are based on principles of learning that emphasize the co-construction of knowledge by learners, who in this case are the teachers themselves. Teachers in the PLC meet regularly to explore their practices and the learning outcomes of their students, analyze their teaching and their students' learning processes, trust each other while sharing their difficulties, draw conclusions, and make changes to improve their teaching and their students' learning (Tschannen-Moran, 2014). The concept of PLC arose in the field of education in the context of workplace-based studies conducted in the 1980s that addressed teachers whose professional relations were characterized by continuous striving for improvement, focused on student learning, and who collaborated and explored their work. Such relationships differ from the norms used in the teaching of a more individualistic culture, which typically characterizes schools as a place of work (Lortie and Clement, 1975).

In 1982, Little conducted an anthropological study of six primary and secondary schools in four counties in the western USA (Little, 1982). He found that schools with norms of collaboration, collegiality, and research could respond better to the pressures of external changes and education initiatives. This finding was reinforced by Rosenholtz (1989), who combined surveys and interviews with 78 primary schools. She distinguished "rich" and "poor" schools with respect to learning. The learning-rich schools were more likely to establish norms of cooperation and continuous improvement.

Newmann (1996) argued that a professional community of teachers offers a supportive environment in which teacher learning can occur. For example, the Center for Organizing and Building in Schools at the University of Wisconsin conducted systematic research on 24 primary, junior high, and high schools in which structural and organizational changes were carried out, with an emphasis on the quality of instruction in mathematics and social sciences. It was found that aspects of a school's professional community that include common norms and values, a focus on student learning, reflective dialog, transformation of teachers' practice in public classes, and a focus on collaboration, are linked to robust teaching and support for teacher learning.

In a series of articles based on analysis of the NELS:88 database, Lee *et al.* (1997) argued that more organized schools produce higher levels of teacher satisfaction, positive student behavior, problem-solving pedagogy, and understanding and learning in mathematics and science:

> Our results indicate that when there is a professional community of teachers – when teachers are taking responsibility for the success of all their students – more than learning is occurring. (Lee *et al.*, 1997, p. 142)

Shulman (1997), in his lecture at the Mandel Institute in Israel, spoke enthusiastically about the idea of both teacher communities and student communities. Shulman argued that because a single teacher can never possess perfect knowledge of pedagogical content, we must continue to create conditions in which a teacher can collaborate with other teachers and be part of a community of teachers facing difficult teaching challenges. In other fields, no one expects a single professional working alone to solve an important problem, because complex, real-world problems require "distributed expertise", *i.e.*, the sharing of highly specialized professionals in dealing with common challenges.

Vescio *et al.* (2008) reviewed 11 studies on how teachers' communities influence student teaching and learning. They found that participation in learning communities influenced teaching practice, such that the teachers became more student-centered. The teaching culture also improved, as the community increased the degree of cooperation among teachers, the teachers' authority, their empowerment, and their ongoing learning. They also found that teachers who participate in communities benefit from this, as reflected in improved achievements over time.

Bryk *et al.* (2010) identified professional communities, along with a work culture oriented toward improvement and access to professional development, with elements of "professional capacity" associated with improvements measured in primary school achievement in Chicago over a period of 6 years in the 1990s. A recent study by Kraft and Papay (2014) reinforced this important insight. These researchers used a measure for the professional environment that was composed of the responses of teachers to a survey in North Carolina combined with a national test in mathematics and elementary school reading. They found that teachers who work in a supportive environment, compared to those who work in a less supportive one, have increased effectiveness over time.

PLC workshops for chemistry teachers were initiated in Israel 2 years ago. These workshops were supported by the Ministry of Education and sponsored by the Trump Foundation, the Weizmann Institute of Science, and the National Center of Chemistry Teachers at the Weizmann Institute. The workshop operates on a cascade model: a leading team of researchers guides a group of teachers who will lead communities of teachers in regional communities close to home (Figure 4.2). The prospective leading teachers meet once a week. They develop activities and pedagogical teaching strategies with the leading team of researchers (a "bottom-up" approach), and implement them in their own classes before they disseminate them among the communities of teachers in their regions.

A major topic of discussion in the PLC workshops for chemistry teachers is the diagnosis of students' ideas and difficulties. The teachers who participate in the project are usually surprised to find that their students have learning difficulties and misconceptions. Therefore, during the workshops, they reflect upon their teaching methods, and discuss how to use different strategies to cope with these difficulties. In addition, they encourage the teachers in the regional PLCs to implement the change and then collect and analyze their students' assignments. Major misconceptions have been

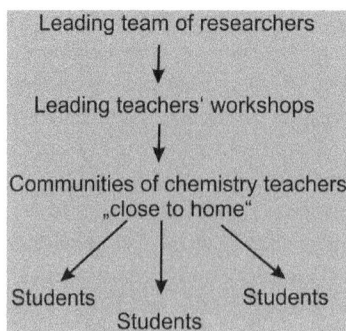

Figure 4.2   The PLC cascade model.

encountered in topics such as Bonding and structure, Acids and bases, Energy, and Equilibrium.

The following example is based on a study conducted by Ben-Zvi *et al.* (1986). The study consisted of three stages: (i) a diagnostic investigation of students' views of structure in chemistry, (ii) development and implementation of a program designed to prevent some of the misconceptions identified in the first stage, and (iii) an evaluation of the new program. The diagnostic investigation of students' views of structure in chemistry consisted of a questionnaire administered to eleven 10th-grade classes in different high schools in Israel (about 300 students, average age 15 years). All students had studied chemistry for at least half a year. The question relevant to the atomic model was (p. 64):

A metallic wire has the following properties: (i) conducts electricity, (ii) brown color, and (iii) malleable. The wire is heated in an evacuated vessel until it evaporates. The resulting gas has the following properties: (iv) pungent odor, (v) yellow color, and (vi) attacks plastics.

1. Suppose that you could isolate one single atom from the metallic wire. Which of the six properties would this atom have?
2. Suppose that you could isolate one single atom from the gas. Which of the six properties would this atom have?

The diagnostic questions were discussed in each regional community, guided by two leading teachers. The teachers in the community that was "close to home" met every other week. The leading teachers initiated the meeting's activities by sharing ideas and learning materials that had been developed at the leading teachers' meetings. Each meeting consisted of:

- an opening activity aimed at creating social and personal relationships among the members of the group, as well as openness and trust to strengthen the cooperation among members of the community, and enable them to gain a sense of ownership
- "our corner": one or two teachers share an experiment or an interesting activity with their colleagues – a short, stimulating and thought-provoking activity that can be applied in the classroom. It can be an experiment, a demonstration, a discussion question, an interesting video clip, or a technological innovation in education
- a discussion referring to a content and pedagogical subject, *e.g.*, diagnostic questions, misconceptions, unclear questions, or alternative assessment methods
- sharing lesson plans regarding new curriculum materials.

The PLC workshops were accompanied by an evaluation study that consisted of questionnaires and interviews. To date, the teachers who have participated in the PLC workshops for chemistry teachers have claimed that the professional community environment improved their self-efficacy and enhanced their ability to share teaching difficulties with their colleagues. They say that during the meetings, a feeling of trust was developed among the participants, which enabled them to discuss and analyze their students' cognitive and affective problems, misconceptions, and learning outcomes. In addition, the fact that they could share ideas, lesson plans, and interesting experiments was an asset in itself.

4.5   Summary

- Encouraging teachers to develop ownership of curriculum innovation is a promising way of actively involving them in the development, adoption, and implementation of the innovations in their practices.
- Involving teachers as curriculum developers and innovators of their own practice is one of the most promising ways for both effective implementation of curriculum change and teacher professional growth.
- Creating professional communities of practitioners is an effective bottom-up way of bringing innovation into the chemistry curriculum and professional development. However, unaccompanied communities of practice face the risk of perpetuating established practices instead of implementing innovations.

References

Barnea N., Dori J. and Hofstein A., (2010), Development and implementation of inquiry-based and computerized laboratories: reforming high school chemistry in Israel, *Chem. Educ. Res. Pract.*, **11**, 218–228.

Ben-Peretz M., (1990), Teachers as curriculum makers, in Husen T. and Postlethwaite N. T. (ed.), *The International Encyclopedia of Education*, Oxford: Pergamon, pp. 6089–6092.

Ben-Zvi R., (1998), *Energy and the Human Being*, Rehovot: Weizmann Institute of Science, in Hebrew.

Ben-Zvi R., (1999), Non-science oriented students and the second law of thermodynamics, *Int. J. Sci. Educ.*, **21**, 1251–1267.

Ben-Zvi R., Eylon B. S. and Silberstein J., (1986), Is an atom of copper malleable? *J. Chem. Educ.*, **63**(1), 64–66.

Bolte C. F., Holbrook J. and Rauch F., (2012), *Inquiry-based Science Education in Europe: Reflections from the PROFILES Project*, Berlin: FU Berlin.

Bryk A. S., Gomez L. M., Grunow A. (2010), *Getting Ideas into Action: Building Networked Improvement Communities in Education*, Stanford: Carnegie Foundation for the Advancement of Teaching, retrieved from www.carnegiefoundation.org/spotlight/webinar-bryk-gomez-building-networkedimprovement-communities-in-education.

Bybee R. W. and Loucks-Horsley S., (2000), Supporting change through professional development, in Resh B. (ed.), *Making Sense of Integrated Science: A Guide for High Schools*, Colorado Springs: BSCS, pp. 41–48.

Cohen D., Ben-Zvi R., Hofstein A. and Rahamimoff R., (2004), On brain, medicines, and drugs: a module for "Science for All" program, *Am. Biol. Teach.*, **66**(1), 9–19.

Conant J. B. and Nash L. K., (1964), *Harvard Case Histories in Experimental Science*, Cambridge: Harvard University.

Fullan M., (1991), *The New Meaning of Educational Change*, New York: Teachers College Press.

Hofstein A., Eilks I. and Bybee R., (2011), Societal issues and their importance for contemporary science education – a pedagogical justification and the state-of-the art in Israel, Germany and the USA, *Int. J. Sci. Math. Educ.*, **9**, 1459–1483.

Hofstein A. and Kesner M., (2006), Industrial chemistry and school chemistry: making chemistry studies more relevant, *Int. J. Sci. Educ.*, **28**, 1017–1039.

Hofstein A., Mamlok R. and Rosenberg O., (2006), Varying instructional methods and students assessment methods in high school chemistry, in McMahon M., Simmons P., Sommers R., DeBaets D. and Crawley F. (ed.), *Assessment in Science*, Arlington: NSTA, pp. 139–148.

Holbrook J., (2005), Making chemistry teaching relevant, *Chem. Educ. Int.*, **6**(1), 1–12.

Kraft M. A. and Papay J. P., (2014), Can professional environments in schools promote teacher development? Explaining heterogeneity in returns to teaching experience, *Educ. Effect. Pol. Anal.*, **36**, 476–500.

Lawrenz F., (2001), Evaluation of teacher leader professional development, in Nesbit C. R., Wallace J. D., Pugalee D. K., Country-Miller A. and DiBiase W. J. (ed.), *Developing Teacher Leaders*, Columbus: ERIC Clearing House.

Lee V. E., Smith J. and Croninger R., (1997), How high school organization influences the equitable distribution of learning in mathematics and science, *Sociol. Educ.*, **70**, 128–150.

Levy Nahum T., Hofstein A., Mamlok-Naaman R. and Bar-Dov Z., (2004), Can final examinations amplify students' misconceptions in chemistry? *Chem. Educ. Res. Pract.*, **5**, 301–325.

Little J., (1982), Norms of collegiality and experimentation: workplace conditions of school success, *Am. Educ. Res. J.*, **19**, 325–340.

Lortie D. C. and Clement D., (1975), *Schoolteacher: A Sociological Study*, Chicago: University of Chicago.

Loucks-Horsley S., Hewson P. W., Love N. and Stiles K., (1998), *Designing Professional Development for Teachers of Science and Mathematics*, Thousand Oaks: Corwin Press.

Loucks-Horsley S., Stiles K. E., Mundry S., Love N. and Hewson P. W., (2010), *Designing Professional Development for Teachers of Science and Mathematics*, 3rd edn, Thousand Oaks: Corwin Press.

Mamlok R., (1998), *Science: An Ever-developing Entity*, Rehovot: Weizmann Institute of Science, in Hebrew.

Mamlok R., Ben-Zvi R., Menis J. and Penick J. E., (2000), Can simple metals be transmuted into gold? Teaching science through a historical approach, *Sci. Educ. Int.*, **11**(3), 33–37.

Mamlok-Naaman R., Hofstein A. and Penick J., (2007), Involving teachers in the STS curricular process: a long-term intensive support framework for science teachers, *J. Sci. Teach. Educ.*, **18**, 497–524.

Newmann F. M., (1996), *Authentic Achievement: Restructuring Schools for Intellectual Quality*, San Francisco: Jossey-Bass.

Norris S. P. and Phillips L. M., (2003), How literacy in its fundamental sense is central to scientific literacy, *Sci. Educ.*, **87**, 224–240.

Obaya O., (2003), Action research: creating a context for science teaching and learning, *Sci. Educ. Int.*, **14**(1), 37–47.

Priesner C., (1991), How the language of chemistry developed, *Chem. Act.*, **33**, 11–21.

Rayner-Canham M. and Rayner-Canham G., (1998), *Women in Chemistry*, Washington: ACS.

Rosenholtz S. J., (1989), *Teachers' Workplace: The Social Organization of Schools*, White Plains: Addison-Wesley Longman Ltd.

Seybold P. G., (1994), Provocative opinion: better mousetraps, expert advice, and the lessons of history, *J. Chem. Educ.*, **71**, 392–399.

Shulman L. S., (1997), *Communities of Learners & Communities of Teachers*, Jerusalem: Mandel Institute.

Stuckey M., Mamlok-Naaman R., Hofstein A. and Eilks I., (2013), The meaning of relevance in science education and its implications to science curriculum, *Stud. Sci. Educ.*, **49**, 1–34.

Tomorrow 98, (1992), *Report of the Superior Committee on Science Mathematics and Technology in Israel*, Jerusalem: Ministry of Education and Culture, English edition: 1994.

Tschannen-Moran M. (2014), *Trust Matters: Leadership for Successful Schools*, John Wiley & Sons.

Vescio V., Ross D. and Adams A., (2008), A review of research on the impact of professional learning communities on teaching practice and student learning, *Teaching and Teacher Education*, **24**, 80–91.

Wellington J., (1991), Newspaper science, school science; friends or enemies? *Int. J. Sci. Educ.*, **13**, 363–372.

Westerlund J. F., Garcia D. M., Koke J. R., Taylor A. T. and Mason D. S., (2002), Summer scientific research for teachers: the experience and its effects, *J. Sci. Teach. Educ.*, **13**, 63–83.

Yarden A., Brill G. and Falk H., (2001), Primary literature as a basis for a high-school biology curriculum, *J. Biol. Educ.*, **35**(4), 190–195.

CHAPTER 5

Action Research as a Philosophy for Chemistry Teachers' Professional Development and Emancipation

This chapter deals with the potential of different forms of action research in the continuous professional development of chemistry teachers. Theoretical resources, namely Grundy's three modes of action research and the Interconnected Model of Teacher Professional Growth (IMTPG), which differentiate and reflect different modes of action research, are discussed. Examples from the USA, Israel, and Germany are presented. One detailed case focuses on building professional competencies in a group of teachers who are reflecting upon and developing their own individual professional practices in Israel. An accompanying university educator assumed the role of training the teachers in potential strategies and methods, and helped them carry out their own individual action research projects. The other detailed example describes an ongoing 20-year cooperation of a group of chemistry teachers from different schools accompanied by an educator from a university in Germany. The central focus of that project is the joint development of new lesson plans for wide dissemination. The chapter closes with a discussion of the potentials of both examples using a joint theoretical framework.

5.1 Every Teacher a Researcher

One set of medium- to long-term strategies for teachers' professional development consists of the wide variety of methods and applications of action research (*e.g.*, Feldman, 1996; Parke and Coble, 1997; Bencze and Hodson,

Advances in Chemistry Education Series No. 1
Professional Development of Chemistry Teachers: Theory and Practice
By Rachel Mamlok-Naaman, Ingo Eilks, George Bodner and Avi Hofstein
© Rachel Mamlok-Naaman, Ingo Eilks, George Bodner and Avi Hofstein 2022
Published by the Royal Society of Chemistry, www.rsc.org

1999; Towns *et al.*, 2000; Eilks and Ralle, 2002). Each of these methods has a slightly different strategy; however, all of them include strong bottom-up, teacher-centered components. Action research is unique as a set of methods for teachers' professional development in that it operates a critical approach by using research-related activities performed by teachers to directly change teaching and learning practices.

According to Feldman (1996), the first goal of action research in educational settings is not to generate new knowledge – whether local or universal – but rather to change and thereby improve classroom practices. Nevertheless, the generation of new knowledge might be considered to have a subsidiary role, depending on the action research mode chosen and on the objectives negotiated within the group of involved participants (Eilks, 2014). In the end, the development of individual practices and the generation of results of general interest can be understood as two sides of the same coin, with both holding equal importance (Eilks and Ralle, 2002). However, both central goals differentiate action research from much of the traditional understanding of educational research.

The first edition of the AERA Handbook of Research on Teaching (Gage, 1963) contained a chapter on "Experimental and quasi-experimental designs for research". This chapter had such a significant effect on the methodology for educational research that it was issued as a separate volume (Campbell and Stanley, 1963). It is tempting to trace the common, albeit erroneous, belief that assessment and evaluation are more or less synonymous back to the popularity of quasi-experimental designs that concentrated exclusively on the difference (if any) between the performance of students in an experimental *versus* control group on a common exam.

Bodner *et al.* (1999) described the fundamental flaw in this erroneous model for the evaluation of new approaches to teaching in terms of a metaphor that they labeled the "sports-mentality approach". All that matters to the casual soccer fan is whether their team won. More committed fans want to know that the final score was 2 to 1. As was noted, however, in the paper in which this metaphor was introduced, "... it is difficult to imagine a coach selecting a team for the next match based only on this information" (Bodner *et al.*, 1999).

Those who believe that there are more useful approaches to monitoring the effects of the changes that we make in our courses might be tempted to invoke a statement attributed to the poet and classicist A. E. Housman who, in 1903, described an individual's work as follows: "...he uses statistics like a drunk uses lampposts, more for support than illumination." Even if we accept the notion that changes in the cognitive domain are particularly important, it is suggested that there are other domains from which useful information for evaluating the effects of changes in instruction can be collected (Bodner, 2016).

More than 30 years ago, for example, Metz (1987) studied the difference between the use of interactive modes of instruction and traditional lecture-based approaches for an introductory, college-level chemistry course in the

USA (Bodner *et al.*, 2014). The comparison started in the traditional cognitive domain by demonstrating that there was no statistically significant difference between student performance in the common exams. The most useful insight, however, came from data collected in the affective domain, where changes in the students' attitudes toward the course were evaluated. With gained experience in the evaluation of curriculum reform projects, it was recognized that it is also useful to look for improvements in students' ability to retain their knowledge or understanding when they move on to the next chemistry course, as well as their ability to transfer their knowledge/ understanding to a course in another department for which the evaluated course is a prerequisite. Moreover, the usefulness of asking whether there are approaches to instruction that can improve students' ability to think like a practicing chemist was recognized (Cartrette and Bodner, 2010; Bhattacharyya and Bodner, 2014).

Before discussing concrete approaches to using action research to improve high-school instruction in chemistry, we explore how action research can be used to achieve the concept of *Every Teacher a Researcher* (Carr and Kemmis, 1986; Kemmis and McTaggart, 1988; McTaggart, 1991). This introductory section therefore focuses on situations in which instructors can apply action research within their own classroom, or by collaborating with a colleague who can help obtain relevant data because of restrictions placed on collecting data in the instructor's own course by an Institutional Review Board (IRB).

The foundation for this approach to action research is a series of assumptions that start with the idea that any significant change in instruction should be accompanied by some form of evaluation. For any substantive change, this evaluation needs to go deeper than asking: "Do students like it?" We advocate thinking about questions such as: "What do students learn that they were not learning before?" and "If we could provide students with a voice to express their opinions and concerns, what changes would they recommend?"

One important characteristic of action research is that everyone in the classroom – students, teachers, and researchers – are active participants in the study. This is not research done *on* teachers and their students, but *with* teachers and their students. To illustrate the difference between what happens in action research and normal classroom practice, we will use an example from a recent workshop. While describing what it means to "give a voice" to students who are seldom heard, the participants were asked to raise their hand if they had involved one or more students from that year's classes in the process of selecting the textbook for the next year. Fewer than 10 hands were raised from among a group that included perhaps 300 college- or university-level science faculty members.

Action researchers assume that the instructor is what Schön (1987) referred to as a "reflective practitioner". Whereas work by Piaget, Dewey, or Lewin emphasized reflection-on-action, Schön advocated reflection-in-action (Schön, 1987). Instructors should be continuously asking themselves: "What do I do?", "How do I do it?", "What does this mean for me as an instructor and for my students?" As patterns in this thought process arise, so does the recognition that there are "problems" that need to be solved. By trying different approaches, talking to colleagues, going to professional meetings, and so on, the instructor comes up with the idea for a particular "plan" for taking some form of "action" to address a particular problem.

The next assumption can be understood by carefully analyzing statements such as: "My students like...". People who understand action research are tempted to respond: "All of them?" (not likely). Whereas traditional approaches to evaluation separate students into experimental *versus* control groups to look for an effect, proponents of action research recognize that any substantive change in instruction will have both positive and negative effects. The fundamental questions are therefore: "Cui bono?" [Who benefits?], and "Who does not?" A particular intervention within the context of action research might therefore be targeted toward a particular group of students.

While the "action" takes place, the group involved in the process "monitors" and "evaluates" what happens by collecting pertinent data. Action research is therefore described as formative (rather than summative), informal, subjective, interpretative, reflective, and an experience-based form of inquiry. When the data are examined for trends, new strategies are developed that are designed to further maximize the potential benefit of the intervention and minimize any harm, and these strategies are fed into the next cycle of plan, act, monitor, and evaluate.

Many academic disciplines differentiate between cyclical and linear forms of time. Individuals and groups sometimes focus on the cyclical nature of time captured by the celebration of the fall and spring equinoxes or a midsummer's night. At other times, they look at life in terms of the progression from birth to childhood and beyond. Figure 5.1 emphasizes a cyclical perspective on action research, which is taken from a description of action research by Hunter (2007). It portrays action research as a spiral that goes forward in time.

In a case study at Purdue University in the USA, the chemical education research group applied action research to a variety of courses, including three projects that involved advanced-level chemistry or chemical

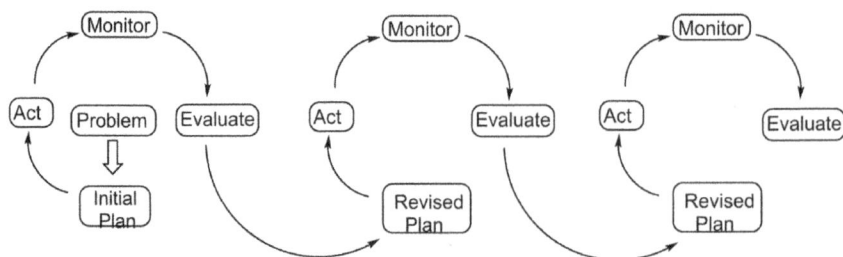

Figure 5.1 An illustration of the cyclical nature of action research.

engineering courses. Although action research methodology can be applied to one's own course, our work was constrained by the need to collect data in the absence of any form of coercion, duress or constraint as defined by federal regulations governing human subject research. A graduate student was therefore involved in each project to facilitate the collection of data that could not be traced back to individual students.

One project involved comparing the use of high-level, workstation-based computer simulations of real-world industrial processes in a capstone chemical engineering course that focused on "design". The goal was to compare the effect of using these computer simulations as an alternative to design exercises that involved the use of equipment that was analogous to one component of a pilot-plant version of an industrial process. In each case, students were responsible for generating a set of optimized design characteristics that would deliver the products specified for that year's class. If we had restricted our analysis to the results of the Likert-scale surveys used as part of this study, we would have reached an erroneous conclusion. By actively involving both the professor and his students in the evaluation process, we were able to address the question: "Which approach was a more 'realistic' experience?" The answer that we obtained was unexpected because both approaches were considered more "realistic" by certain students in the course. The students argued that the computer simulation was more "realistic" because it was a large-scale simulation of an industrial process and most of their time could therefore be devoted to the design process, rather than collecting data in the laboratory. But they also argued that traditional experiments were more "realistic" in the sense that they had to develop an understanding of the effect of manipulating various combinations of valves, controls, and so on. As a group, the participants in the class concluded that the optimum approach to this course was to have three 5-week projects, two using traditional equipment and one using one of the computer simulations.

Another project was carried out in a second-year organic chemistry course taught in our College of Pharmacy at Purdue University in the USA. The action involved replacing traditional lectures with a problem-oriented approach in which students worked in groups of three to answer problems introduced by the instructor, who then helped them examine the logical consequences of their answers. To encourage students to learn how to work together, the groups were allowed to discuss the questions on each exam before they separated to write their individual responses. Although members of a group often wrote different answers to the exam questions, the class as a whole demonstrated a better understanding of the exam questions than ever before. As noted elsewhere (Bodner *et al.*, 1999):

> The action research methodology helped to answer questions that might not otherwise have been asked, such as: How do we overcome student resistance to this approach? What do we have to do to ensure that groups operate effectively? What is the nature of the dissatisfaction that might lead an instructor to change to a problem-oriented approach? What

factors make it difficult to change the classroom environment? What factors interfere with the ease with which this technique can be used by other instructors, or transported to other institutions? What effect does this mode of instruction have on the instructor's attitude toward teaching? What effect does it have on students' perception of the difficulty of organic chemistry? Does this approach to instruction produce students who are more likely to think the way an organic chemist thinks?

This course represented an unusual example of action research because the action research spiral extended over a period of more than 10 years as different aspects of the classroom environment were altered.

Holladay (2001) summarized her work as chronicling "a three-year effort to improve the classroom experience for both students and their instructor in an analytical [chemistry] course for non-majors". This course was selected for intervention because of a general perception that the students intensely disliked it, but no one knew why. The process of applying action research techniques to this course began by identifying problems from the students' point of view. Their concerns were collected, ideas were proposed to address those concerns, and changes were then implemented in the second year. As some concerns were addressed, others surfaced that generated other proposals for changes in the course. The solutions implemented over the duration of this project ranged from simple changes that could be implemented quickly to others that were complex and required long-term effort to incorporate them into the course. Significant changes were made in the content of both the lecture and laboratory components of the course so that it more closely matched the students' interests and needs. It is interesting to note that the students advocated increasing the level of the material covered in lectures. Furthermore, the instructor reported that the improvements made in the course had a positive effect on his enjoyment of teaching it.

The key features of the use of an action research methodology in these examples can be summarized as follows: it facilitates the process of making changes in either what we teach or how we teach it; evaluation occurs while the changes are being made; as many sources of information are collected as possible; both quantitative and qualitative data are found to be useful; we never presume that all students will benefit from a given change; we are constantly searching for ways to maximize positive effects and minimize negative effects of the changes being made; and the students are knowing, active participants in the decision-making process concerning the changes that should be made in the next iteration of the innovation cycle (Bodner *et al.*, 1999).

5.2 Action Research in Chemistry Education for Teachers' Professional Development

As we can see from the previous discussion, there are many different ways of using action research in science education in general, and chemistry education in particular. Laudonia *et al.* (2017) recently presented a systematic review of approaches to the use of action research.

Differentiation among the different strategies of action research is often based upon: (i) the type of researcher–practitioner relationship, and (ii) the degree to which the approach appears to implement preplanned concepts efficiently *versus* encouraging practitioners to ask their questions and develop their personal practices based on their own needs (Eilks and Ralle, 2002). Within the wide range of possibilities, action research can be operated either as a practitioner-oriented inquiry into teachers' work and their students' learning in the classroom, or for the development of new teaching strategies for widespread dissemination oriented to teachers' and students' deficits or personal interests (Eilks, 2014).

All of the approaches to using action research in chemistry education consider it to be a cyclical process of planning, implementation, observation, and reflection (Figure 5.1). Implementing changes and improving classroom practices are considered to be iterative processes within authentic practice. Feldman and Minstrel (2000) argue that this is one of the main reasons why action research should become an integral part of converting theoretical change into actual practice. It reduces the generational time lag between the acquisition of new knowledge and its application in class. Action research cycles of reflecting on and changing authentic practice allow teachers and researchers alike to evaluate classroom procedures for ongoing improvement. Action research enables teachers to develop more finely-tuned teaching materials and curricula in a step-by-step fashion (Eilks and Ralle, 2002), but it also allows teachers to better understand their field of practice, which can contribute to their professional development and growth (Eilks and Markic, 2011).

The critical and emancipatory philosophy behind action research separates it from more conventional research and professional development strategies (Wadsworth, 1998). Based on large amounts of evidence from science education and beyond, action research has become a widely accepted tool for teachers' professional development at all stages of their careers, including pre-service training (*e.g.*, Gore and Zeichner, 1991). Loucks-Horsley *et al.* (1998) claimed that action research's main strength as a professional development strategy is the fact that teachers define the research questions, contribute to broader research interests in a meaningful way, and are actively and directly involved in the research process. The use of action research as a strategy for professional development is based upon the following assumption:

Teachers are intelligent, inquiring individuals with important expertise and experiences that are central to the improvement of education practice. By contributing to or formulating their own questions and by collecting data to answer these questions, teachers grow professionally. Teachers are motivated to use more effective practices when they are continuously investigating the results of their action in the classroom. (Loucks-Horsley *et al.*, 1998, p. 97)

For action research to be an effective means of helping teachers reflect on their personal teaching practices, we must provide them with opportunities to engage in life-long professional development (Eilks and Markic, 2011). These opportunities should offer an environment of support, cooperativeness, and collaboration with either educational researchers or other teachers teaching the same or related subjects, which is particularly important when traditional modes of support are remote and mediated solely by digital communication technologies (Laudonia and Eilks, 2018). Ideally, this must be an environment that encourages self-reflection on personal classroom practices and the results of their research efforts. According to Holly (1991), collaboration can be viewed as a major form of professional development. When teachers jointly reflect upon their own experiences, they have the opportunity to critically examine themselves, improve their teaching ability, and better understand their students' learning difficulties. This can be profoundly useful for the development of new, potentially more effective teaching and learning scenarios (Obaya, 2003).

5.3 Different Models of Action Research for Chemistry Teachers' Professional Development

Action research has a long tradition, dating back to the 1940s, in general (Roth, 2007) and in science education in particular (Laudonia *et al.*, 2017). A variety of models that apply action research are available for chemistry education. Their range covers everything from teacher-centered research with the core objective of improving own individual practice, to research-centered projects built around more general questions. Examples also include developing effective and motivational teaching scenarios, which will ideally have the potential to guide the whole community of teachers who are interested in innovations in a particular subject (Eilks, 2014).

The most common criterion for differentiating between the various forms of action research is the role played by the teacher, especially with regard to exactly who maintains control in an action research project (Grundy, 1982). Beginning as a rather research-oriented variant in its early stages, action research then morphed into teacher-centered versions, which became firmly established in the educational arena by the 1970s (Altrichter and Gstettner, 1993). Consequently, Grundy (1982) and Carr and Kemmis (1986) differentiated between three different modes of action research: technical, practical (also called interactive), and emancipatory (see Figure 5.2). Whereas emancipatory approaches of action research seem to offer the most promise for helping teachers reduce existing deficiencies in their individual practices, Eilks and Ralle (2002) argued for using action research to focus on more general questions of educational research. When applied to general questions, the objective of the study should be the wide dissemination of new curricula and teaching methods. In this context, a more research-oriented mode seems to be the most appropriate. This does not mean, however, that

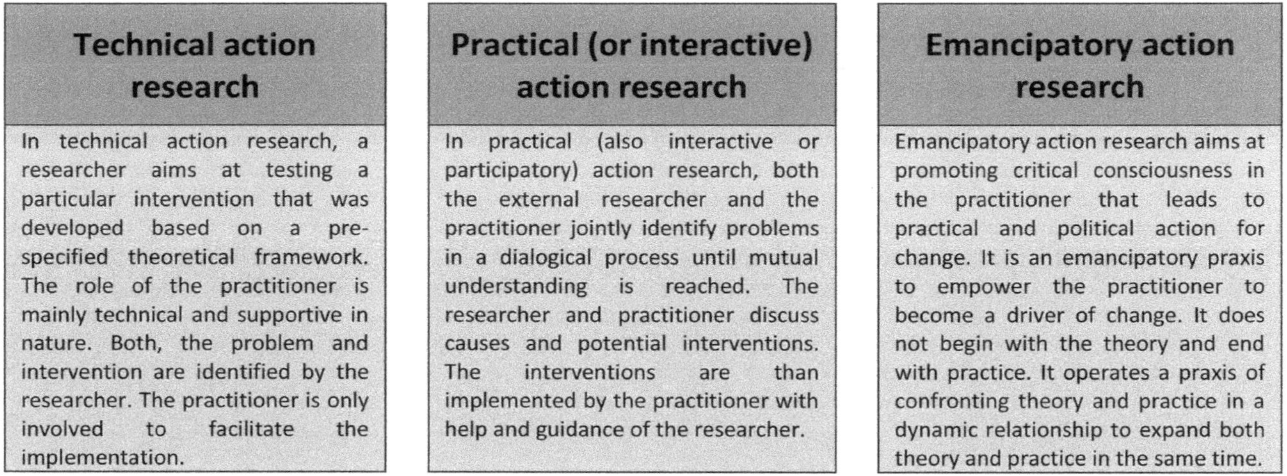

Technical action research	Practical (or interactive) action research	Emancipatory action research
In technical action research, a researcher aims at testing a particular intervention that was developed based on a pre-specified theoretical framework. The role of the practitioner is mainly technical and supportive in nature. Both, the problem and intervention are identified by the researcher. The practitioner is only involved to facilitate the implementation.	In practical (also interactive or participatory) action research, both the external researcher and the practitioner jointly identify problems in a dialogical process until mutual understanding is reached. The researcher and practitioner discuss causes and potential interventions. The interventions are than implemented by the practitioner with help and guidance of the researcher.	Emancipatory action research aims at promoting critical consciousness in the practitioner that leads to practical and political action for change. It is an emancipatory praxis to empower the practitioner to become a driver of change. It does not begin with the theory and end with practice. It operates a praxis of confronting theory and practice in a dynamic relationship to expand both theory and practice in the same time.

Figure 5.2 The three modes of action research based on Grundy (1982) and Masters (1995).

the role of the teacher in university-driven action research should ever be reduced to purely technical support.

As already noted, Grundy (1982) discussed three basic modes of action research that are differentiated by the overall degree of the practitioner's personal responsibility. This specifically includes the question of whether teachers must make important content-based, methodological, and research-based decisions, or whether such decisions are left to the accompanying researchers, or perhaps have been made by the accompanying researchers before the start of the cooperation with the teachers.

We can use Grundy's (1982) classification as a structural model for differentiating between different types of action research projects. In the case described by Eilks and Markic (2011) and Eilks (2014), however, Grundy's categories seem to lend themselves to a model of professional development. This was also predicted by Grundy (1982, p. 363), who stated that a move between the different modes is also possible:

> The differences in the relationship between the participants and the source and scope of the guiding 'idea' can be traced to the question of power. In technical action research it is the 'idea' which is the source of power for action and since the 'idea' often resides with the facilitator, it is the facilitator who controls power in the project. In practical action research the power is shared between groups of equal participants, but the emphasis is upon individual power of action. Power in emancipatory action research resides wholly within the group, not with the facilitator and not with an individual within the group. It is often the change in power relationships within a group that causes a shift from one mode to another.

In the experiences described in the project by Eilks and Markic (2011) and Eilks (2014), the several years of interactive, close cooperation began with an organization best described as technical. This maloperation of the interactive mode in the first year was caused by low self-confidence among the teachers and initially lacking competencies when it came to applying jointly negotiated methodological approaches. But the teachers' self-esteem grew rapidly in proportion to their burgeoning capabilities in the other areas. Consequently, stronger participatory processes and a subsequent emancipation from the established framework – including external authorities – came to pass (Figure 5.3).

Grundy's (1982) model and its connection to the theories of participation and emancipation (Figure 5.3) offer a framework for reflecting upon cases of action research presented in subsequent sections of this chapter. In this reflection, the chapter focuses on the potential of action research to allow authentic professional development by teachers. One of the most recognized models for reflecting on cases as a form of teachers' professional development is the Interconnected Model of Teacher Professional Growth (IMTPG) by Clarke and Hollingsworth (2002), which is more broadly elaborated upon

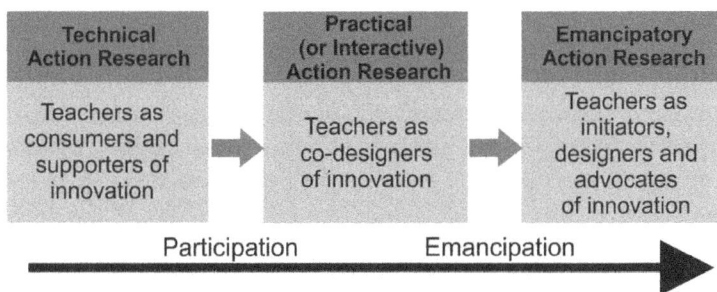

Figure 5.3   Professional and personal development progressing through an ongoing change through different modes of action research (Eilks and Markic, 2011).

in Chapter 2 of this book. The IMTPG describes teachers' professional development in four domains: (i) the personal domain (beliefs, attitudes, and previous experiences), (ii) the practical domain (the teacher's authentic teaching practices), (iii) the external domain (topic requirements, media, and curriculum aspects), and (iv) the domain of consequences (goals and effects). Each of the IMTPG domains has subdomains that can be differentiated in terms of content knowledge (CK), pedagogical knowledge (PK), and pedagogical content knowledge (PCK), introduced by Shulman (1986). The combination of the IMTPG and Shulman's professional knowledge domains (detailed in Chapter 2 of this book) form the basis for understanding the concept of teachers' professional development by action research in the two cases described herein.

5.4   Exemplifying Different Forms of Action Research in Chemistry Education

5.4.1   A Professional Development Course Facilitating Teachers' Application of Action Research in Israeli Chemistry Education

This project (*i.e.*, Dass *et al.*, 2008) focused primarily on allowing teachers to develop their own individual practices by enabling them to conduct small-scale action research studies in their schools. It began with a workshop structure that was created to build the teachers' confidence in their ability to conduct action research as part of their professional development. The workshop was held as part of a wider educational program for chemistry teachers. Action research was incorporated into this program to: (i) provide teachers with a powerful tool for enhancing their professional expertise by performing small-scale research projects in their local environments, (ii) improve opportunities to practice this technique in the teachers' schools, and (iii) create a professional community of connected, collaborating chemistry teachers.

In planning the program, it was assumed that the teachers needed to improve their CK, PCK, and leadership skills. Action research was introduced to offer potential solutions for achieving all of these goals using a joint approach. The action research segment of the program was structured around a series of workshops teaching the methodology and research tools necessary for data collection and evaluation. This was carried out by coupling workshop phases with action research activities performed in the teachers' regular school environments. In between the workshop sessions, the teachers were asked to discuss the content of the workshop with their school colleagues and to apply the learned strategies and methods to aspects of their own practice, including specific research involving different domains: pedagogical, affective, behavioral, and cognitive.

A total of 22 teachers participated in the program. All participants were experienced secondary-school teachers (Mamlok-Naaman *et al.*, 2003). Each person had at least 10 years of experience teaching high-school chemistry (grades 10–12), including preparing their students for matriculation examinations. All of the teachers had at least a BSc in Chemistry; 12 held higher academic degrees as well. In addition, several of the teachers had backgrounds in academic science research. The teachers participating in the workshop series had been recommended by their principals and regional tutors. They had also been classified as highly motivated teachers, who would be able to implement changes in the way chemistry is taught in their schools. The program involved weekly meetings and consisted of a total of 450 hours at the National Center for Chemistry Teachers, located in the Science Teaching Department of the Weizmann Institute of Science. The action research portion of the course consisted of eight meetings focusing on, among other things, development of the teachers' PCK. Planning of the workshops had been based on the assumptions that the participants were experienced teachers with backgrounds in both chemistry and pedagogy, but lacked experience in performing educational research and were not familiar with qualitative research paradigms in general, and action research in particular.

The action research sessions dealt specifically with: (i) action research principles, (ii) the qualitative research approach, (iii) methodology (the rationale for choosing a subject for research, defining good research questions, types of research tools and data collection, methods of data analysis), (iv) self-reflection during each of the stages, and (v) the presentation of reports. During the first four meetings, the workshop leaders presented the theoretical framework and issues. Subsequent meetings were devoted to discussing the various stages of the action research steps as operated in this project (Figure 5.4).

After each meeting, the participants were asked to meet with their colleagues at school and share the topics and subjects discussed in the workshops. The goal here was to involve the teachers' entire chemistry staff in the action research process. During the course meetings, participants reported on their work, discussed the difficulties that had arisen from their teamwork, and received comments, clarification, and support from other participants,

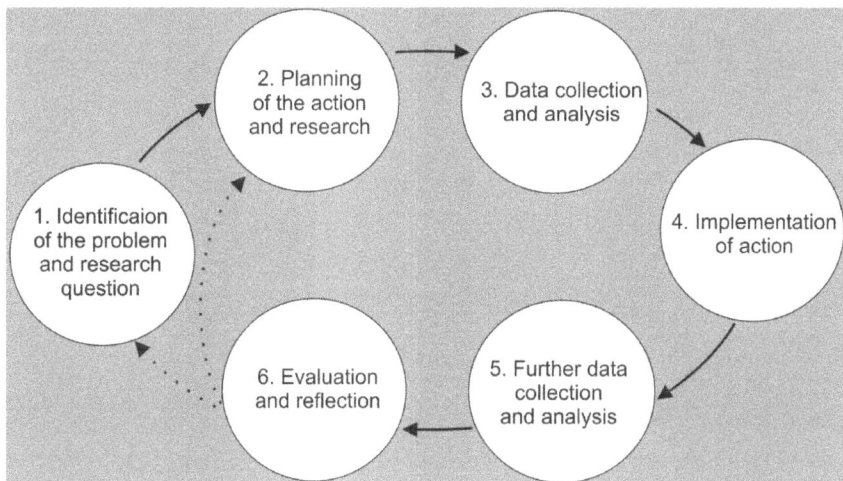

Figure 5.4 The cyclical approach of action research as operated in the project.

and from the facilitators. The participants were asked to choose research questions relevant to their work in school and in their classrooms. The questions formulated by the teachers covered various aspects of their environment. Some teachers were interested in their students' cognitive understanding. Others were interested in pupils' behavior or motivation. Several of the teachers were interested in their own teaching and how it functioned in the classroom. Each of the teachers focused on a different topic, thus enabling the workshop participants to be involved in a wide range of research issues.

The workshop was accompanied by a study of its effects (Mamlok-Naaman *et al.*, 2004, 2005). The main goal of the study was to discover how the action research workshop influenced the participants' professional development. Ten female teachers who addressed our call volunteered to participate in this study. Three sources of data were used for evaluation of the workshop: (i) a Likert-scale attitude questionnaire administered after the course, (ii) follow-up interviews (30 min each) employing open-ended questions given 1 year later, and (iii) self-reported stories provided by the participating teachers. The questionnaire was administered after completion of the workshop. The four-point Likert scale items (with 4 = "fully agree" and 1 = "do not agree") included seven items assessing the teachers' opinions regarding the extent to which the workshop had contributed to both their professional development and their ability to continue using action research methods in their teaching practice. Table 5.1 lists the seven items in the questionnaire and the mean and standard deviation for the teachers' responses.

Most of the teachers expressed satisfaction with the workshop. This was particularly true with respect to their personal interest in conducting action research and in becoming part of a community of practice. Some participants did not feel that the workshop had broadened their teaching strategies.

Table 5.1 Teachers' attitudes regarding how the action research workshop contributed to their work ($n = 10$).

Statements related to the action research workshop	x	SD
It increased my interest in integrating action research into my own class	3.80	0.42
It encouraged me to strengthen my relationship with science teaching experts	3.70	0.48
It improved my teaching strategies	2.80	0.92
It improved my ability to reflect upon my work	3.70	0.48
I became part of a community of practice	3.80	0.42
I would be happy to participate in a continuing workshop on action research	3.60	0.52
I would recommend that my friends participate in a similar workshop	3.90	0.32

However, because our candidates were all experienced teachers, we assume that most of them already possessed a large repertoire of such strategies.

Three categories emerged from the interviews when a grounded theory approach was applied to the interview data (Glaser and Strauss, 1967): (i) enactment followed by reflection, (ii) membership in a community of practice, and (iii) contact with science teaching experts. In the analysis, 7 of the 10 participants stressed the fact that they had learned the importance of reflecting upon their work using the methods associated with the action research process. Some of them conducted interviews with their students, stating that their pupils raised important points that contributed to their work, as can be seen in the following quotes:

> The tools we received during the workshop were important for me; I have been using them in my classroom. I will definitely continue to use them, and I know for sure that the new strategy I implemented is now a part of my teaching. My work improved and I have a new way of teaching and dealing with problems in the classroom and reflecting upon my work.

> I had a good feeling about the action research project I prepared during the workshop. This year I dealt with the same problem again with this tool. I performed all of the research cycles without even calling this a project and I am happy with the students' grades. I used action research to cope with the students' difficulties.

Eight out of ten teachers claimed that they felt like members of a team during the workshop meetings and that they continued to exchange ideas over a year later. One of the teachers said:

> During the workshop meetings we consulted with each other; you cannot be a part of a group and not work. I met other people who had the same problems that I was facing. We could share opinions and help each other. We worked together all the time and were very active. We kept up contact

through emails and exchanged information and ideas. For instance, the discussions in the workshop helped me define my action research question. I presented my project to the group and I received meaningful feedback. It was wonderful for me.

During the action research workshop, the teachers established closer contact with the academic staff who were involved in the workshop on a professional basis. They could contact the experts and consult with them about problems whenever they needed to:

I think that without the close contact and the instruction of the co-ordinators, I would not have been able to conduct an action research study. I was a science researcher before, and I had to change the way I think. It took me some time to understand the research methodology, but I developed it thanks to the workshop.

One example of self-reporting deserves particular attention. Lia wrote that she had begun teaching a 10th-grade chemistry class in a new school. The working atmosphere and overall environment in the new class greatly disturbed her. When asked to choose a subject for her action research project, Lia described the cooperation problems in her 10th-grade class of 28 students. She wanted to enhance her pupils' ability to collaborate with one another and openly share ideas during the lessons, in order to improve the learning environment, by integrating computer work into her lesson plans, while also introducing group work. Accordingly, her research question asked: "What impact does integration through group work and the use of technological learning tools have on the classroom environment?" Lia divided her students into small groups and added computers as part of her teaching strategy. This enabled each group to progress at its own speed, using a variety of auxiliary materials, including software, web quests, Internet sites, and various literature sources. As research tools, Lia incorporated pre- and post-lesson questionnaires and classroom observations, as well as holding informal discussions with her students. She also discussed the process with her colleagues and advisors during the workshop. From her students' reactions, Lia concluded that the action research workshop had directly influenced her work and teaching. She reported that these changes had become an integral part of her teaching practice.

The follow-up study revealed that Lia's process of change included two additional phases: she worked closely with the teaching team at her school, sharing the principles of action research with her colleagues. She also became head of a project called "Learning from Success" that was based on reflective practice. Her goal was to identify problems in school and find project-based solutions that were perceived as successes. Together with the teaching team, she integrated action research principles as a tool for implementing changes.

Analysis of the interviews and questionnaires revealed that the action research workshop had helped the participants to both reflect on their

personal practices and make appropriate changes. More specifically, teachers started investigating their own work and systematically exploring how
their students were learning subject matter as a core element of developing
personal PCK. Action research proved itself to be a new and rewarding experience for this group of teachers.

A further objective of the workshop was to enhance the chances of creating
a professional community of chemistry teachers. It had been suggested that
teachers need to have a strong and solid professional foundation in order for
them to develop socially (Hofstein *et al.*, 2003). The participants in our group
had many opportunities to enhance their social skills through collaboration
and cooperation with their peers. The workshop sessions enabled them to
share ideas, consult with each other, and exchange information with the
others as often as they wished, maintaining good social and professional
relations. The cooperation among the teachers in the group was fruitful and
helped promote their teaching strategies. It became a major developmental
source of their knowledge about teaching in the sense intended by Appleton
and Kindt (1999), but in this instance appeared in a reflected form. The
participants reported that they had become a team of active, professional
teachers. In fact, they maintained contact and exchanged information with
one another in the year after completion of the workshop. They even met a
few extra times and worked together. From the beginning, the approach
chosen for this study was believed to promote emancipatory action research
as described by Grundy (1982). The creation of such independent learning
communities among teachers is a clear indicator that Grundy's processes of
participation and emancipation, as mentioned in his three-mode developmental model of action research, had begun to take root.

Another central objective of the workshop was to establish a team of lead
teachers who would perform action research with other colleagues as the
ultimate stage of the emancipatory process. The action research project
described here enabled the participants to develop and enhance their own
social abilities, and to gather experience as lead figures. Such leaders serve
as catalysts for change and as supporters of reforms (Hofstein, 2001). This
action research program helped our teachers develop the necessary leadership skills. The participants reported direct sharing of the methodology of
action research with their colleagues.

We can conclude that the program presented here fulfilled its objectives.
By integrating action research into their work, the teachers learned to better
understand both their students and how their students think and learn.
They also increased their professional esteem and directly shared their
experiences with their colleagues. Both critical reflection upon their own
experiences and exchanges with trusted colleagues provided a promising
environment for strengthening the teachers' professional repertoire.
Professional development in the sense of improved leadership skills and
increased networking (both teacher–teacher and teacher–researcher interactions) seems to offer promising glimpses into increasing teacher participation in the professional community. It may also contribute to teachers'

emancipation within the educational system in the sense of being willing and able to self-reflect upon exactly when and why one should involve external experts in one's own teaching practices. All four domains of the IMTPG were touched upon in the process. The personal and practical domains were changed by a growing body of own and shared experiences. This included changes in beliefs and attitudes, which stemmed from participants' authentic practices and a better way to arrive at effective methods. The development of altered teaching settings also added to the effectiveness of the approach. The remaining domains (external and consequences) were also influenced, not only by curricular changes but also by the rethinking of the relationship between practitioners and external partners. Part of this entailed a more critical reflection upon external fostering/hindering factors that affect the educational system.

5.4.2 A 20-Year Curriculum Development Project for German Chemistry Education Based on Participatory Action Research

This project (*i.e.*, Eilks, 2003, 2014; Eilks and Markic, 2011) was started by a group of six chemistry teachers from different secondary schools, accompanied by a chemistry educator from the university. It has been under way since the summer of 1999. The start of the project was also inspired by a debate initiated by De Jong in several papers and conferences, *e.g.*, De Jong (2000), in which he argued that substantial curriculum development at the school level can only be realized through a give-and-take process based on teaching experience and existing research evidence. This process acknowledges that both empirically validated research results and experientially based teacher knowledge form the two ends of a spectrum of teaching and learning knowledge that are equally important and have their own strengths and weaknesses (McIntyre, 2005).

In the beginning, the focus of the group's work was on the development, testing, and evaluation of alternate teaching approaches on the particulate nature of matter. The focus later shifted to implementation of methods for cooperative learning under inclusion of modern information and communication technology, and finally to developing and researching socio-scientific-based issues and education for sustainable development approaches to chemistry teaching. After 2004, several members of the community of practice characterized by this project also began to spread the overall implementation of these alternate approaches by authoring a new textbook series for schools in cooperation with the university researcher.

This project has been based on participatory action research (Whyte *et al.*, 1989), specifically the interpretation adopted for education research as described by Eilks and Ralle (2002). Participatory action research, as described in Figure 5.5, seeks to improve teaching practices through the close cooperation of university-level science education researchers and in-service

New concepts and materials for teaching	Knowledge about teaching and learning	Developed practice	Trained teachers	Documentation of teaching practice

Development of teaching strategies and materials

Testing in practice

Aims:
Concepts and knowledge for the development of teaching practice
Development of concrete practice by the research process

Reflection and revision

Evaluation

Field of teaching practice

Knowledge about learning processes	Teaching experiences	Didactical and methodological reflections	Scientific back-ground simplified for education	Teachers' intuition and creativity

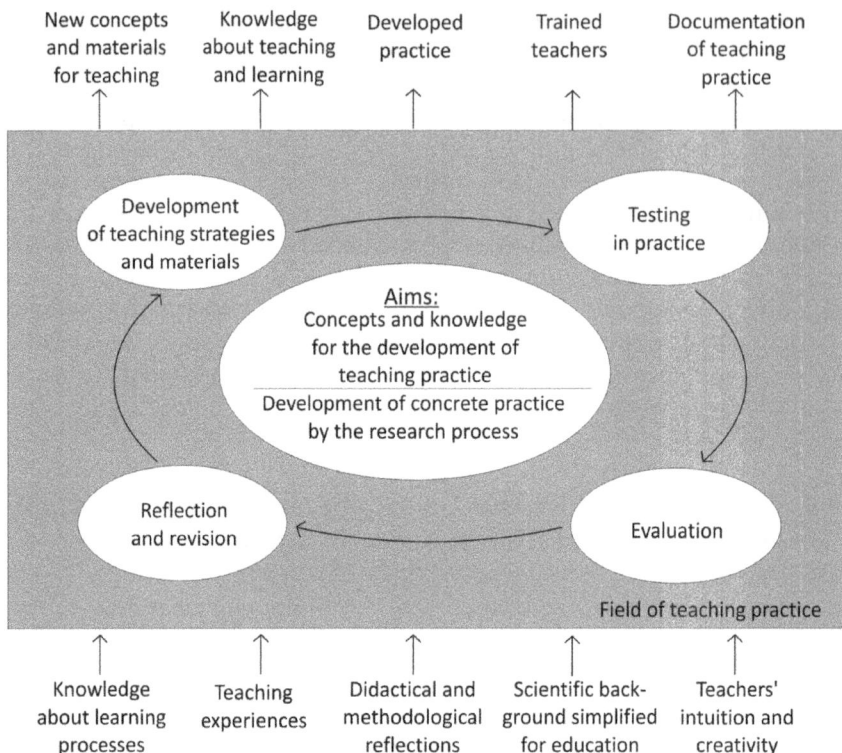

Figure 5.5 Participatory action research in science education.

teachers. It seeks to develop new curricular and methodological approaches and analyze them in authentic teaching situations, thus leading to an evidence-based understanding of the effects of newly developed teaching approaches. The model also aims to make sustainable changes in the fields touched by these innovations and seeks contribution of the professional development of all involved participants.

To achieve such research-based innovation, a cyclical model of brainstorming, evaluation, reflection, and revision is applied. Any ideas for classroom innovation are continually compared to the evidence available from empirical research. To connect these two areas, relevant research evidence is presented to the teachers by the university researcher in a group discussion format. Empirical results are compared to actual teaching experiences in the classroom and examined with respect to the needs and wishes expressed by teachers for their day-to-day situation in school.

The original work group, which included six chemistry teachers, has expanded to a current total of 15 participants from various types of secondary schools. The current members have widely varying professional qualifications, ranging from just 2 to over 30 years of teaching experience. Since the third year of the project, 10 to 12 group members have taken a regular and active part in the monthly meetings. Four teachers in this core group have been involved since the project's inception in 1999, four more joined within the first year, and the remainder became involved in the second and third project years. Another three teachers were invited to join the group in 2016 because the first teacher from the initial group retired, although he still attends project meetings. Several other teachers are loosely associated with the group, but come to the group's meetings infrequently due to longer travel distances. The same university educator has accompanied the group throughout the 20 years of the project. In addition to the core members, student teachers and doctoral students from the accompanying university are involved with the group from time to time.

In the first 7 years of the project, minutes were taken during the group's regular monthly meetings. In addition, group discussions were conducted at least once yearly to monitor any potential changes in teachers' views and the working process. For the first few years, the group discussions tended to focus on the teachers' self-esteem and reflect upon their development over the years. The relationship between the practitioners and researchers was also a focus of the group discussions. Each group discussion lasted roughly 60 minutes and was recorded in video and audio format. Data were analyzed by qualitative content analysis (Mayring, 2000). Group discussions were specifically evaluated with respect to changes in individual teachers' attitudes and their own estimations of the changes occurring in the practical aspects of their teaching. From the fourth year onward, a growing saturation of changes in teachers' attitudes was observed. After the sixth year, the focus of the group discussions systematically shifted toward questions of potential future activities, implementation, and dissemination (Eilks and Markic, 2011).

Over the years, a continuous shift in the teachers' attitudes and views on practice-to-research relations could be observed (Eilks, 2003, 2014; Eilks and Markic, 2011). In the first year, the teachers viewed themselves mainly as technical supporters of innovation as described by Grundy's (1982) technical mode (Figure 5.2). This stemmed from uncertainty as to the level of trustworthiness and security in the newly developed curricular and methodological approaches. Nevertheless, after the initial year, all of the teachers expressed the feeling that the new approaches had proven themselves to be better than the old ones. They saw definite advantages in the new approaches that could be recognized in their changed practices. One teacher explicitly addressed the differences perceived in accepting suggestions brought into teaching practice *via* ordinary channels:

> What I really liked was that we were getting input from teachers who stand in the classroom every day on the one hand, and on science education research, from the researchers, on the other. I have my copies of science education journals at home and leaf through them when I have the time, but quite honestly I lack the time to translate them into teaching concepts. [The researchers] can really look at what is happening overall with methodology in Germany as a whole.

Even at the end of the first year, it was very apparent that the point of view of the practitioners had begun to change, *e.g.*, contributions in teachers' journals were being seen "with a different view". Evidence of a developing attitude of "personal ownership" became repeatedly clear, even as a growing distance toward certain authorities (*e.g.*, authors of curriculum materials in teachers' journals) emerged. With regard to the teachers' personal development, one other idea started to dominate, namely the exchange of ideas about personal teaching practices among the group members. The participants stated that "simply [...] speaking with the others [...] and getting ideas [...] automatically raised the level of professionalism".

The discussion after the second year showed an even stronger shift toward self-reflection among the teachers. Instead of mainly debating the advantages and disadvantages of the changed teaching approaches, the group started reflecting upon the meaning of the process for the individual. All of the participants noted a clear increase in their own activity inside the group. This was often connected to starting a second trial run of the altered approaches with the fresh possibility of receiving feedback and carrying out joint reflections. Statements like "Now I know what I should have done differently last year" were typical of this mindset. The teachers also said that the long-term cooperation had led to increasing openness inside the group and a tendency to self-confidently and proactively bring their own criticisms and ideas into play. The teachers described an increased feeling of being able to competently reflect upon their own teaching and an ability to better exchange ideas. Three teachers in the group explicitly described themselves as becoming increasingly reflective and self-critical toward their own previous teaching practices. One participant even defined the main value of working in a group as combating any tendency to become "pedagogically fossilized by years of teaching".

The development of reflective competence went hand-in-hand with the teachers' growing ownership of the curricular and methodological approaches that they had changed. Two teachers described this change as leading "from a teacher who initially wanted to be trained to a convinced defender of the new concept" or "from being a consumer in a group to an activist". The participants themselves expressed further changes due to their own professionalization, with a focus on "a totally different view toward methodological variety" as the main point. The teachers named the exchanges taking place inside the group, the contact with science education research findings, and comparison of these results with their own beliefs and experience as sources of the changes that occurred.

From year 2 onwards, individual group members started their own initiatives for the group, with an even bigger effort seen in the third project year. They also became active in initiating in-service training seminars in their own schools and in the environment surrounding their schools. First attempts were also made to transfer the changed teaching approaches and methods to other topics and/or subject areas.

The central focus in the group discussion after the third year changed to a growth in self-confidence toward being a leading teacher for educational change and in the teachers becoming able to change practice with an increasing independence, or emancipation, from the accompanying science educator. The teachers named this a "change of roles". The competence of the supervisor and the possibilities represented by university support were still considered integral and necessary. Determination of the approach and/or main content emphasis by the external chemistry educator, however, barely occurred after the third year. Another growing focus in the fourth to sixth years was an intensified discussion of the role of stakeholders in the educational arena.

Whereas in the early years of the project, discussion always touched upon the question of whether the new ideas fitted the governmental syllabus and the regulations set up by the schools, in subsequent years the discussion shifted. The teachers now described a growing distance between themselves and authors of teachers' journals and textbooks, including educational policy authorities, who suggested curricula and teaching approaches without providing any evidence of their effects. A growing distance from the traditionalist approach of "copy–pasting" old syllabi into new ones was mentioned. There was increasing criticism of every textbook being "a compromise of the lowest common level". The teachers clearly demanded more freedom in operating within the governmental regulations "allowing for more openness and innovation". They felt freer and more self-confident in stretching the regulations set up by governmental authorities or within the school when implementing their self-developed, student-oriented chemistry teaching practices. In this year, some of the teachers enthusiastically accepted the offer of becoming members of a team to implement and widely disseminate their work and ideas by writing a new school chemistry textbook.

Despite the intensive cooperation, this project remained primarily in the realm of technical action research during the first year, even though it was originally planned to be practical or interactive in the sense of Grundy's (1982) ideas on participatory action research. The cooperative structure changed dramatically from the second year onward to the originally intended interactive and participatory mode.

The systematic build-up of a process fostering the development of equal roles as envisioned by both Altrichter and Gstettner (1993) and the participatory action research model by Eilks and Ralle (2002) was increasingly noticeable during this period of the project. As suggested by Noffke (1994) and Dickson and Green (2001), this facilitated dismantling the obstacles and hierarchical attitudes existing between participants and researchers at the start of the project. Even during the switch to this interactive, participatory mode, clear signs of nascent teacher emancipation became clear, at least when dealing with authorities from outside the immediate group (textbook authors, writers for teachers' journals, governmental regulators). This trend became especially strong after the third year. Nevertheless, the teachers still requested guidance in the form of the external expertise offered by the participating university educator. However, this could now be viewed as a thoughtful decision on the part of the teachers. Perhaps our participants' ability to make informed decisions about when to follow or oppose so-called experts on a topic, including the aspect of self-reflection before selecting or rejecting which people to follow on a given issue, can be seen as one of our successful contributions to teacher emancipation.

The project demonstrated that attitudes toward innovation can change dramatically when introduced in an interactive and participatory action research approach. When teachers are involved in long-term innovative research and are given equal rights as described above, their attitudes and abilities undergo a positive change with respect to testing and implementing new ideas. This leads to teacher-based innovations stemming from their own convictions in the sense of a constructive rethinking of their ideas. Their PCK changes permanently in all five areas described by Magnusson *et al.* (1999; see Chapter 2) including: (i) a common orientation with regard to teaching, (ii) increased reflection on conscious changes that can be made to the original curriculum and further developed, (iii) improvements in the way evaluation improves the teachers' knowledge inspection, (iv) expansion of teachers' knowledge about learners as a result of dealing with empirical teaching/learning research and personal self-reflection, and (v) development of strategies for conveying concepts to others due to the creation of new teaching/learning environments.

Although many of the innovations introduced by teachers in this project stemmed from what they learned as the group discussed the science education literature, the switch from technical to participatory to emancipatory action research mirrors the central elements of teachers' emancipation. Even now, after almost 20 years of cooperative work, the teachers have purposely avoided emancipation in the sense of total independence from university support. The understanding that both teachers and science educators have equal footing, albeit different roles, is now deeply engrained on both sides. The cooperation and simultaneous participation of both parties in their professional and teaching practices has led to the desire for intensified contact between schools and universities for quite simple and pragmatic reasons: access to information and resources and an understanding of the different, but complementary, types of expertise possessed by the researchers and practitioners in the project (McIntyre, 2005).

All four domains of the IMTPG (Clarke and Hollingsworth, 2002) also played an important role in both the overall work and the teachers' progress in their professional development within this project. In particular, the elements related to the IMTPG involved the use of: (i) foreknowledge and needs of the participants as input for development in the personal domain, (ii) systematic input from empirical research in the external domain as teachers expanded their knowledge and expertise in these fields, (iii) new approaches to authentic classroom situations in the domain of practical relevance that led to more well-reflected experiences among teachers, and

(iv) sustainably implemented leadership coupled with reflection upon the effects of changed practices due to the evaluation steps within the context of the domain of consequences.

5.4.3 Comparing the Two Cases

We reflected on examples of applying action research to chemistry education that started from different perspectives (Mamlok-Naaman and Eilks, 2012). One example was an interactive course design, which successfully qualified experienced teachers to use action research in innovating and reflecting upon new practices in their own individual school environments. It also taught them the necessary skills to become innovative leaders with the ability to inform their school colleagues and to implement action research beyond their own classrooms. The second project created a network of teachers from different schools, accompanied by a university science educator. This group learned how to develop and research innovative curricular and methodological approaches in chemistry education with the goal of generating both curricular structures and teaching materials that could be widely disseminated. It also contributed to changing individual teaching practices and to promoting teachers' ongoing professional development *via* the action research process. Both projects can be considered successful in their own ways (Mamlok-Naaman and Eilks, 2012). The advantage of the first case lay in the workshop structure. The necessary contact time with the accompanying researcher and the overall duration of the project were both limited and could therefore be repeatedly applied to different groups of teachers in close succession, allowing the process to reach a relatively large number of teachers and schools. The project demonstrated that the action research strategies and methods to which the teachers were exposed had been successfully applied. The teachers' feedback also showed that they considered the program to be helpful. Two of the limiting factors in this example might be the relative lack of widespread dissemination of individual findings to the scientific community at large and limitations in the teachers' ongoing systematic access to newly gained evidence from conventional research. In this respect, the second research project showed considerably more potential than the first. Due to the ongoing, long-term cooperation, the teachers have received continuous input from the research side. The results could also be widely disseminated in various arenas, up to and including school textbooks to be marketed throughout the country. Moreover, changes within the entire chemistry curriculum were – and still are – developed and implemented. However, the main limitation of this model is the intense, long-term cooperation with the university researcher. Although this approach can achieve long-lasting, deep penetration in a specifically chosen area, its depth and intensity are offset by the limited number of practitioners who can effectively be accompanied by one university researcher.

Both projects can also be considered successful when it comes to aiding teachers' professional development (Mamlok-Naaman and Eilks, 2012). In the first case, the participants achieved higher levels of professionalism by taking ownership of new strategies for better reflecting upon and improving their teaching practices. It is hoped that this will make them better able to cope with their own practices and help contribute to their future development. Self-reflection and contemplation, as shown by the example of Lia above, clearly provide evidence for this claim. In the second case, the resultant group discussions evidenced changed attitudes and personal capabilities in the field of curriculum innovation. The teachers reported learning gains in the personal (beliefs, attitudes, and previous experience) and practical (teacher's authentic teaching practices) domains of the IMTPG. They also expressed changed personal opinions in the external domain (topic requirements, media, and curriculum) and in the domain of consequences (goals and effects). In the case of the Israeli participants, connections to the IMTPG were more difficult to discern due to the stress placed upon teachers following their own, extremely diverse interests during the action research studies. However, we can see that taking part in the studies touched on the participants' personal domain. Changes in coping with practice are also a part of teachers' learning in the practical domain. We can further assume that the teachers' external domain and domain of consequences were influenced indirectly. In any case, these areas can be affected by an individually applied, small-scale action research project touching upon the proper topics regardless of the IMTPG's assumption that all four domains are influenced whenever changes take place in any one of them.

Thus, both examples share joint objectives that can be considered to have been successfully accomplished. It was clear from the beginning that both projects were aimed at teachers' professional development as discussed above. The learning goals of each approach were successful with respect to the different dimensions of PCK in both cases. Positive effects were clearly demonstrated in all four PCK domains in both examples. At least in the German case, explicit mention was made of the positive impact of the action research experience on the teachers' views of educational research and its output. Such a change can be seen as potentially beneficial for closing the current gap that exists between theory and research in science education (Loughran, 2007). Therefore, actively supporting the IMTPG using structural elements borrowed from PCK represents a good combination in terms of action research's ability to successfully improve science teachers' knowledge.

However, the essential issue of applying action research as an expression of critical theory in educational settings (Kemmis, 1993) lies beyond the generation of knowledge and professional learning. This issue focuses on action research's successful contribution to teacher emancipation in the sense of Grundy's (1982) third mode of action research. Even in this area, both cases showed indications of progress. The first case motivated teachers within an interactive course design to apply action research principles to become multiplicative factors for implementation of action research and, consequently, leaders of innovation in their own schools. The second example aided teachers' growth through a process that can be characterized using Grundy's different modes of action research as a developmental model (Grundy, 1982). The teachers started as technical supporters of innovation, developed themselves through an interactive process of curriculum development, and finally emancipated themselves more and more in the sense of applying and advocating alternate teaching strategies in settings outside of the cooperative research project.

The core idea to be discussed here is not the successful application of Grundy's (1982) emancipatory mode of action research. It is the emancipatory effect of the activity itself, which has shown potential as a quality indicator for successful teachers' professional development in chemistry education in the sense of action research. The indicators of teachers' professional emancipation were even more important than the clear signs that essential IMTPG elements were present in both case studies. The teachers in both projects became leaders, either acting as multiplicative factors of action research methodology in the first example, or actively becoming textbook authors or in-service trainers disseminating their action research results in the second. The teachers in both situations increased their competence with respect to critically reflecting upon both the framework and the circumstances in which they have to work. They also became more open to taking the initiative, including a willingness to change and develop their professional practices with respect to their own needs. Instead of "professional development", this might be well characterized by the definition of "professional growth" used in the IMTPG. This might be the most powerful, long-reaching contribution that action research can offer to chemistry teachers' professional development.

5.5 Summary

- The action research strategy helps teachers be researchers in their own practice. It aims to develop concrete teaching practices in a cyclical approach of identifying deficits, implementing changes, and researching and understanding corresponding effects for further improvement.
- Action research for chemistry teachers' professional development can follow a range of models – from accompanying teachers in single chemistry education action research case studies, to involving teachers in long-term collaborative projects of curriculum and practice development.
- Action research is one of the most emancipatory and powerful strategies for teachers' professional development because it enables the teachers to understand and implement change in their classrooms based on research evidence.

References

Altrichter H. and Gstettner P., (1993), Action research: a closed chapter in the history of German school science, *Educ. Act. Res.*, **1**, 325–360.

Appleton K. and Kindt T., (1999), How do beginning elementary teachers cope with science? Development of pedagogical content knowledge in science, Paper presented at the annual meeting of NARST, Boston, USA.

Bencze L. and Hodson D., (1999), Changing practice by changing practice: toward more authentic science and science curriculum development, *J. Res. Sci. Teach.*, **36**, 521–539.

Bhattacharyya G. and Bodner G. M., (2014), Culturing reality: how organic chemistry graduate students develop into practitioners, *J. Res. Sci. Teach.*, **51**, 694–713.

Bodner G. M., (2016), Changing how data are collected can change what we learn from discipline-based educational research, in Jones D. L., Ding L. and Traxler A. (ed.), *Proceedings of the 2016 Physics Education Research Conference*, College Park, USA.

Bodner G. M., MacIsaac D. and White S. R., (1999), Action research: overcoming the sports mentality approach to assessment/evaluation, *Univ. Chem. Educ.*, **3**(1), 31–36.

Bodner G. M., Metz P. A. and Casey K. L., (2014), Twenty-five years of experience with interactive instruction in chemistry, in Devetak I. and Glažar S. A. (ed.), *Learning with Understanding in the Chemistry Classroom*, Dordrecht: Springer, pp. 129–148.

Campbell D. T. and Stanley J. C., (1963), *Experimental and Quasi-experimental Designs for Research*, Chicago: Rand McNally.

Carr W. and Kemmis S., (1986), *Becoming Critical: Education, Knowledge and Action Research*, London: Falmer.

Cartrette D. P. and Bodner G. M., (2010), Non-mathematical problem solving in organic chemistry, *J. Res. Sci. Teach.*, **47**, 643–660.

Clarke D. and Hollingsworth H., (2002), Elaborating a model of teacher professional growth, *Teaching and Teacher Education*, **18**, 947–967.

Dass P., Hofstein A., Mamlok R., Dawkins K. and Pennick J., (2008), Action research as professional development of science teachers, in Erickson I. V. (ed.), *Science Education in the 21st Century*, Hauppauge: Nova, pp. 205–240.

De Jong O., (2000), Crossing the borders: chemical education research and teaching practice, *Univ. Chem. Educ.*, **4**(1), 29–32.

Dickson G. and Green K. L., (2001), The external researcher in participatory action research, *Educ. Act. Res.*, **9**, 243–260.

Eilks I., (2003), Co-operative curriculum development in a project of participatory action research within chemical education: teachers' reflections, *Sci. Educ. Int.*, **14**(4), 41–49.

Eilks I., (2014), Action research in science education: from a general justification to a specific model in practice, in Stern T., Rauch F., Schuster A. and Townsend A. (ed.), *Action Research, Innovation and Change*, London: Routledge, pp. 156–176.

Eilks I. and Markic S., (2011), Effects of a long-term participatory action research project on science teachers' professional development, *Eurasia J. Math. Sci. Techn. Educ.*, **7**(3), 149–160.

Eilks I. and Ralle B., (2002), Participatory action research in chemical education, in Ralle B. and Eilks I. (ed.), *Research in Chemical Education – What Does This Mean?* Aachen: Shaker, pp. 87–98.

Feldman A., (1996), Enhancing the practice of physics teachers: mechanisms for the generation and sharing of knowledge and understanding in collaborative action research, *J. Res. Sci. Teach.*, **33**, 513–540.

Feldman A. and Minstrel J., (2000), Action research as a research methodology for study of teaching and learning science, in Kelly A. E. and Lesh R. A. (ed.), *Handbook of Research Design in Mathematics and Science Education*, Mahwah: Lawrence Erlbaum, pp. 429–455.

Gage N. L. (1963), *Handbook of Research on Teaching*, Washington: AERA.

Glaser B. G. and Strauss A. L., (1967), *The Discovery of Grounded Theory: Strategies for Qualitative Research*, Hawthorne: Aldine.

Gore J. and Zeichner K., (1991), Action research and reflective teaching in preservice teacher education: a case study from the United States, *Teaching and Teacher Education*, 7, 119–136.

Grundy S., (1982), Three modes of action research, *Curr. Persp.*, **2**(3), 23–34.

Hofstein A., (2001), *Action research: involving classroom-related studies and professional development studies*, Paper presented at the IOSTE conference, April 29–May 2, Paralimni, Cyprus.

Hofstein A., Carmi M. and Ben-Zvi R., (2003), The development of leadership among chemistry teachers in Israel, *Int. J. Sci. Math. Educ.*, **1**, 39–65.

Holladay S. R. H., (2001), Action research as the vehicle for curriculum change in analytical chemistry: a longitudinal study, Ph.D. Dissertation, Purdue University, West Lafayette, USA.

Holly P., (1991), Action research: the missing link in the creation of schools as centers of inquiry, in Lieberman A. and Miller L. (ed.), *Staff Development for Education in the 90's: New Demands, New Realities, New Perspectives*, New York: Teachers College Press, pp. 133–157.

Hunter W. J., (2007), Action research as a framework for science education research, in Bodner G. M. and Orgill M. K. (ed.), *Theoretical Frameworks for Research in Chemistry/Science Education*, New York: Prentice Hall, pp. 152–171.

Kemmis S., (1993), Action research and social movement: a challenge for policy research, *Education Policy Analysis Archives*, 1, retrieved from http://epaa.asu.edu/epaa/abs1.html.

Kemmis S. and McTaggart R., (1988), *The Action Research Planner*, Geelong: Deakin University Press.

Laudonia I. and Eilks I., (2018), Teacher-centred action research in a remote participatory environment – a reflection on a case of chemistry curriculum innovation in a Swiss vocational school, in Calder J. and Foletta J. (ed.), *(Participatory) Action Research: Principles, Approaches and Applications*, Hauppauge: Nova, pp. 215–231.

Laudonia I., Mamlok-Naaman R., Abels S. and Eilks I., (2017), Action research in science education – an analytical review of the literature, *Educ. Act. Res.*, advance article.

Loucks-Horsley S., Hewson P. W., Love N. and Stiles K. E., (1998), *Designing Professional Development for Teachers of Science and Mathematics*, Thousand Oaks: Corwin.

Loughran J. J., (2007), Science teacher as learner, in Abell S. K. and Lederman N. G. (ed.), *Handbook of Research on Science Education*, Mahwah: Lawrence Erlbaum, pp. 1043–1066.

Magnusson S., Krajcik J. and Borko H., (1999), Nature, source, and development of pedagogical content knowledge, in Gess-Newsome J. and Lederman N. G. (ed.), *Examining Pedagogical Content Knowledge*, Dordrecht: Kluwer, pp. 95–132.

Mamlok-Naaman R. and Eilks I., (2012), Action research to promote chemistry teachers' professional development – cases and experiences from Israel and Germany, *Int. J. Sci. Math. Educ.*, **10**, 581–610.

Mamlok-Naaman R., Navon O., Carmeli R. and Hofstein A., (2003), Teachers research their students' understanding of electrical conductivity, *Austr. J. Educ. Chem.*, **62**, 13–20.

Mamlok-Naaman R., Navon O., Carmeli R. and Hofstein A., (2004), A follow-up study of an action research workshop, in Ralle B. and Eilks I. (ed.), *Quality in Practice Oriented Research in Science Education*, Aachen: Shaker, pp. 63–72.

Mamlok-Naaman R., Navon O., Carmeli M. and Hofstein A., (2005), Chemistry teachers research their own work two case studies, in Boersma K. M., De Jong O. and Eijkelhof H. (ed.), *Research and the Quality of Science Education*, Heidelberg: Springer, pp. 141–156.

Masters J., (1995), The history of action research, in Hughes I. (ed.), *Action Research Electronic Reader*, Sidney: The University of Sidney, retrieved from www.docstoc.com/docs/2187576/THEHISTORY-OF-ACTION-RESEARCH.

Mayring P., (2000), Qualitative content analysis. Forum: Qualitative Social Research, **1**, retrieved from www.qualitative-research.net/fqs.

McIntyre D., (2005), Bridging the gap between research and practice, *Cambridge J. Educ.*, **35**, 357–382.

McTaggart R., (1991), *Action Research, a Short Modern History*, Geelong: Deakin University Press.

Metz P. A., (1987), *The Effect of interactive instruction and lectures on the achievement and attitudes of chemistry students*, Ph.D. Dissertation, Purdue University, West Lafayette, USA.

Noffke S., (1994), Action research: towards the next generation, *Educ. Act. Res.*, **2**, 9–21.

Obaya O., (2003), Action research: creating a context for science teaching and learning, *Sci. Educ. Int.*, **14**(1), 37–47.

Parke H. M. and Coble C. R., (1997), Teachers designing curriculum as professional development: a model for transformational science teaching, *J. Res. Sci. Teach.*, **34**, 773–790.

Roth K. J., (2007), Science teachers as researchers, in Abell S. K. and Lederman N. G. (ed.), *Handbook of Research on Science Education*, Mahwah: Lawrence Erlbaum, pp. 1203–1260.

Schön D. A., (1987), *Educating the Reflective Practitioner: Toward a New Design for Teaching and Learning in the Professions*, San Francisco: Jossey-Bass.

Shulman L. S., (1986), Those who understand: knowledge growth in teaching, *Educ. Res.*, **15**(2), 4–14.

Towns M. H., Kreke K. and Fields A., (2000), An action research project: student perspectives on small-group learning in chemistry, *J. Chem. Educ.*, 77, 111–115.

Wadsworth Y., (1998), *What is participatory action research? Action Research International (Paper 2)*, retrieved from www.scu.edu.au/schools/gcm/ar/ari/arihome.html.

Whyte W. F., Greenwood D. J. and Lazes P., (1989), Participatory action research, *Am. Behav. Sci.*, **32**, 513–551.

CHAPTER 6

Teacher Professional Development for Society, Sustainability, and Relevant Chemistry Education

Chemistry plays a central role in many of the challenges faced by society. Solutions to issues of energy supply, nutrition, mobility, agriculture, healthcare, and many more are related to chemistry applications. The central importance of chemistry for our contemporary society and its sustainable development justifies every citizen's need for some basic understanding of chemistry to be a responsible citizen and active participant in society. This chapter discusses issues related to the "chemistry for all" approach and the necessity to incorporate societal views on chemistry into chemistry teaching and chemistry teacher education.

6.1   The Roots of Society-oriented Secondary Chemistry Education

In the 1980s, "Science for All" (Harms and Yager, 1981; Fensham, 1985) emerged as a new paradigm in science education that presented a new challenge for science educators, both at the developmental level and in the implementation stages of the curriculum (*e.g.*, Eilks *et al.*, 2013). From the beginning of this paradigm shift in thinking about the students for whom our courses should be designed, Harms and Yager (1981) stated that science for all should be part of the education of all who would be "future citizens" – in other words, for all of our students. In their report *Project*

Advances in Chemistry Education Series No. 1
Professional Development of Chemistry Teachers: Theory and Practice
By Rachel Mamlok-Naaman, Ingo Eilks, George Bodner and Avi Hofstein
© Rachel Mamlok-Naaman, Ingo Eilks, George Bodner and Avi Hofstein 2022
Published by the Royal Society of Chemistry, www.rsc.org

Synthesis, they considered four interrelated "goal clusters" for teaching science: (i) personal needs, (ii) societal issues, (iii) career awareness, and (iv) academic preparation. These goal clusters required multiple approaches to science teaching in its authentic context (Yager, 1996). These clusters directly overlap with the three dimensions of relevant science education identified by Stuckey *et al.* (2013), namely individual, societal, and vocational relevance (Figure 6.1). Unfortunately, the societal dimension is too often neglected (Hofstein *et al.*, 2011).

One approach to addressing the question of how to teach "science for all" courses was the development of science–technology–society (STS) approaches to science teaching. STS attempts to present science together with its technological and social manifestations. Yager (1996) claimed that this approach has great potential to enhance attainment of the above-mentioned goals, thereby shaping the character of scientifically literate citizens. The goal of the STS movement was a society in which individuals would be able to make informed decisions about current problems and issues of scientific origin, and act responsibly as a result of those decisions. Proponents of STS approaches to science argued that citizens who understand how science, technology, and society mutually interact will be able to use their knowledge to handle the issues with which they are confronted. The different approaches to the social, economic, and environmental aspects of science

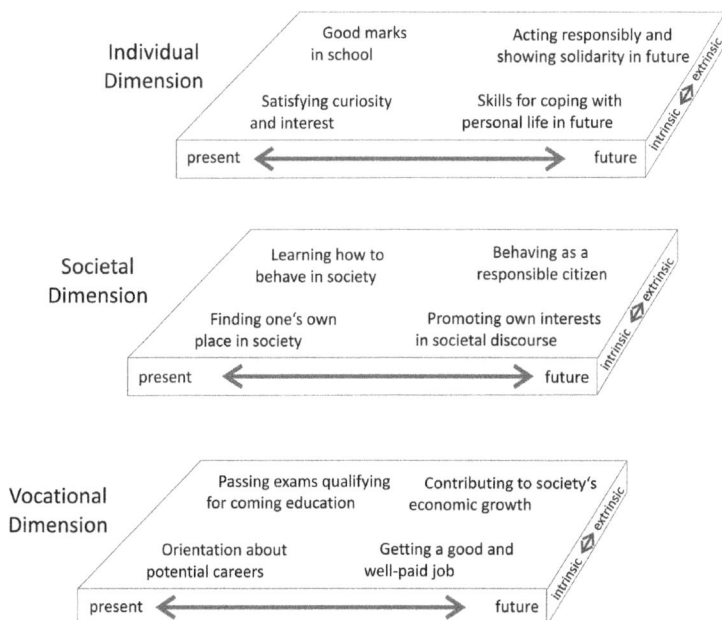

Figure 6.1   The three dimensions of relevant science education and example aspects of relevance allocated to the span of present–future and intrinsic–extrinsic ranges.

represented a fundamental shift in the goals of science courses from traditional curricula that were primarily focused on the acquisition of scientific knowledge. By learning from the perspective of the STS context, students would be taught about natural phenomena in a way that linked science to the technological and societal world in which they live (Hofstein *et al.*, 2011).

To effectively implement STS curriculum materials into chemistry education, we must consider the preparation and professional development of chemistry teachers. The interdisciplinary nature of the subject matter included in the STS programs is demanding (Penick, 1984). Studies have also shown that teachers' beliefs influence the STS implementation process (Tobin *et al.*, 1994). The beliefs that teachers bring to a professional development program can influence the success of a reform initiative. Because teachers are the "agents of change" in educational reform, their beliefs must not be ignored. As others have noted, these beliefs are at the "core of educational change" (Haney *et al.*, 2002).

In investigating STS curricula, Mitchener and Anderson (1987) studied 14 teachers and 200 students. Analysis of the data collected in that study gave rise to three categories: acceptance, rejection, or alteration of the STS curriculum. They found that teachers who felt the STS curriculum enhances their students' motivation to cope with real-life situations improved their attitude to the teaching of STS lessons. However, those who disliked the inclusion of social studies content and the resulting removal of certain science topics rejected the STS reform.

In 2001, Sweeney conducted a study of the effect of incorporating STS issues into science teacher pre-service education courses. In that study, Sweeney (2001) found that prospective elementary- and secondary-school teachers often resisted the incorporation of STS issues as a legitimate component of courses focusing on science teaching methods. Research has also shown that implementation of a wide spectrum of instructional techniques in the science classroom necessitates matching an appropriate assessment tool for each learning goal to measure the students' achievements and progress (Hofstein *et al.*, 1997).

By the 1990s, research suggested that for reform in science education to be successful, extensive, dynamic, and long-term professional development of the teachers needs to take place (Loucks-Horsley and Matsumoto, 1999). Teachers need to receive guidance and support throughout the implementation stages when changes in the curriculum are made (see Chapters 2 and 4 of this book). It is difficult for teachers to make significant changes in either the content or the way they teach. However, teachers are generally excellent learners, and are often interested in trying to teach a new curriculum, as well as in improving and enriching their understanding of "best practices" in pedagogy (Joyce and Showers, 1983).

Eilks *et al.* (2013) studied how teachers implemented an integrated chemistry curriculum based on STS and related approaches, such as education for sustainable development, which differed from the traditional science curriculum. Science teachers usually receive the preparation they need to teach the traditional science curriculum, but they are not prepared to teach science integrated with issues of social science, economy, or the environment. As the curricula evolve, however, teachers need to learn the knowledge, skills, attitudes, and teaching skills that are required to teach interdisciplinary topics found in STS-type science lessons (Bybee and Loucks-Horsley, 2000). They therefore should be encouraged to expand their repertoire of student-assessment strategies to include such techniques as observation checklists, portfolios, and rubrics (Wiggins and McTighe, 1998).

6.2   Learning about Sustainable Development and Green Chemistry

Sustainable development was defined in 1987 as "development that meets the needs of the present without compromising the ability of future generations to meet their own needs" (WCED, 1987). In subsequent years, sustainable development became a guiding principle of international policy. As an essential step toward sustainability, the idea of education for sustainable development (ESD) was suggested (UNCED, 1992). ESD is a skill-oriented educational paradigm aimed at the preparation of young people to become responsible future citizens that is particularly relevant for chemistry education (Burmeister *et al.*, 2012). For sustainable development of our present and future society, decisions need to be made on a variety of topics, such as resource consumption, environmental protection, and new and alternative technologies. Many of these decisions are closely related to chemistry and the chemical industry (Matlin *et al.*, 2015).

Investment in chemistry curriculum development and teacher education is needed to develop general and domain-specific knowledge and skills that will enable students to assess and make decisions about chemistry-based products, technologies, and processes (Burmeister *et al.*, 2012). All students need to develop corresponding skills regardless of whether they embark on a career in science and technology, because they will all be asked to act as responsible citizens in the future and to contribute to societal debates and democratic decision making (Hofstein *et al.*, 2011).

Burmeister *et al.* (2012) suggested four basic ways to integrate ESD into chemistry education: (i) applying Green Chemistry principles to practical work in the school sciences, (ii) using sustainability issues to contextualize chemistry content learning, (iii) addressing technological and environmental challenges in socio-scientific issue (SSI)-based science education, and (iv) innovating school life along sustainability principles. Burmeister *et al.* (2012) suggested SSI-based science education as one of the most promising strategies for the development of domain-specific educational skills for participation in societal debate and decision making about chemistry-related issues of sustainable development.

Stolz *et al.* (2013) suggested criteria for selecting and reflecting societal contexts based on their potential to act as high-quality SSI-based science teaching:

- Authenticity: the topic is authentic because it is currently being openly debated by society. Common media can be checked for proof of authenticity by the presence of the topic in newspapers, magazines, TV, advertising, or social media.
- Relevance: the topic is relevant if respective decisions will affect the current or future lives of the students. Different scenarios are discussed to determine the impact of potential decisions on behavioral choices, consumer behavior, and the availability of consumer products.
- Evaluation in a socio-scientific respect: the SSI can provide for different points of view that can be evaluated from the perspective of science. Media can be analyzed to find situations for which controversial viewpoints are represented by different interest groups, the media, politicians, or scientists.
- Open discussions: the topic must be able to be discussed in an open forum. Arguments need to be tested to make sure that no individuals, religious groups, or ethnic groups will feel insulted or pushed to the fringes of society by its use.
- Based on science and technology issues: the topic must concern itself with a techno-scientific query. Discourse in the media can be analyzed for examples in which scientific facts and concepts are addressed and either explicitly or implicitly used for argumentation.

Many chemistry-related issues of sustainable development are aligned with these criteria. For example, the SSI of using alternative fuels is often discussed in the mass media. Regulations on their use will potentially have a significant impact on the students' future consumer choices. The use of alternative fuels is still controversial, and this issue raises questions regarding chemistry and technology, as well as encouraging public debates. The following examples can serve as a pattern for incorporating SSI into the chemistry classroom.

6.3 Traditional *vs* Alternative Fuels: An Example of Issues of Sustainable Development in the SSI Chemistry Classroom

Issues of sustainable development have been suggested as a way to contextualize chemistry learning for relevant chemistry education (Eilks and Hofstein, 2014). If this is implemented from an SSI-based perspective (Eilks *et al.*, 2013), controversial issues from the sustainability debate can be used to motivate chemistry learning within the context of a societal perspective. The issue of alternative fuels can be used as an example (Mamlok-Naaman *et al.*, 2015).

In recent years, a group of teachers in Israel developed a lesson plan that was called "Can used oil be the next-generation fuel?" (Ezra *et al.*, 2012). This lesson plan focuses students' learning on traditional and alternative fuel sources. The students learn about the advantages and disadvantages of each of the different suggested technologies: fuels from crude oil, recycling of used oils, or producing biodiesel from vegetable oil.

The lesson plan uses a structure that starts with the SSI, involves learning about the content behind the issue, and then turns to questions of evaluation and reflection on the issue from different perspectives and in the foreground of the societal discourse. The lesson plan starts with exposing students to information about the world's energy crisis and its consequences. Discussion of this information activates prior knowledge and raises questions to be answered. The idea that teachers should convey to their students is that sustainable mobility is a worldwide problem and not just a scenario for the science classroom. Furthermore, there are several proposed solutions to this challenge, but these solutions often introduce new problems.

Students undertake different activities to investigate and compare the different fuel types in order to decide on various options for providing fuels for mobility. In one activity, the students are asked to inquire into the chemistry of the use of different fuel types, one of which is biodiesel. Comparative activities require students to select criteria such as enthalpy of combustion values or the release of emissions. The teacher then introduces the student to an experiment that compares the energy released by the combustion of different fuel types. By measuring the mass of the fuel needed to increase the temperature of a certain volume of water by 30 degrees Celsius, students can compare the calorific values of different fuels. They can also investigate the level of pollutants emitted from the burning fuels with a special board called the "Ringelmann scale", which determines the concentration of soot particles produced by the flame. Students are then asked to decide which is the best fuel. Before making a final decision, there is an attempt to involve students emotionally and from an ethical perspective by creating a conflict regarding the use of biodiesel. This activity is based on viewing pictures that highlight the use of crops for fuel instead of using them as a food source for the world's ever-growing population. Decisions made by students should be based on arguments, but first there should be agreement within the group about the assumed meaning of the term "best fuel". This discussion leads to understanding that a thorough comparison requires more criteria beyond the limits of chemical behavior. These criteria include price, environmental behavior, production methods, and societal impact. An open discussion about which technology has the most potential for sustainable development is used to end this lesson plan.

Within this lesson plan, the students learn about an authentic sustainability issue and the complexity of its solution. On the one hand, they learn that there is no "best fuel", or any "best solution" to many sustainability problems. On the other hand, they learn that making use of used oil or biofuels is not "the ideal solution". Other ways might better protect the environment because less waste is produced and fossil resources are saved. However, the students also learn how complex such evaluations are and how many dimensions need to be taken into consideration before an overall decision can be made.

In Germany, Eilks (2002) suggested a similar lesson plan on biodiesel usage. Several years later, another lesson plan was published that focused on the societal debate over the use of bioethanol (Feierabend and Eilks, 2011). Both lesson plans started with authentic examples from the media of public debate. In the case of biodiesel, advertising and brochures from pressure groups were used to set the context of the lesson plan and to provoke questions. In the bioethanol example, articles from news magazines on the concurrency of food and fuel production were analyzed. In both cases, questions were then derived from the material. Student questions regularly concern the science behind the issue. Thus, starting with the public media allows for other questions to arise. These questions encompass aspects of consumer behavior, but also implications of the new technology. These implications are used to construct questions that arise from the different fields of economy, ecology, and society – the three most prominent dimensions in current models of sustainability (Burmeister *et al.*, 2012).

After starting from authentic media, the applied curriculum model (Marks and Eilks, 2009) suggests a phase of clarifying the basic chemistry behind the various issues. This includes the chemistry of fat and transesterification, and alcohols and fermentation. The science background is learned through a combination of theoretical learning and practical work. Experiments are performed that are similar to those suggested in the previous example from Israel. In addition, biofuels are produced by transesterification of rapeseed oil and distillation from fermented grapes.

Reflection on the learned chemistry content clearly indicates that chemistry is important but that it can only help us understand the technology behind the issue. Balanced evaluation and decision making therefore needs to include ecological, economic, and societal aspects. To learn about how society handles SSI, the curriculum model applied in this example suggested the importance of mimicking an authentic societal practice of communication and decision making (Marks and Eilks, 2009; Marks *et al.*, 2014). Consequently, role-playing activities were implemented in both lesson plans. In the biodiesel example, the students mimicked a public panel discussion. The students were the discussants, representing stakeholders from the crude oil industry, a traditional gasoline company, a pressure group for biodiesel promotion, and an environmental protection agency. For the bioethanol example, role playing occurred in the context of a business group. This scenario assumed that a government committee has to come to a decision about whether to make 10% bioethanol in all gasoline compulsory. This is an authentic scenario because Germany was discussing exactly such a law when the lesson plan was developed. In the business game, an imaginary commission is conducting a hearing with different student groups representing chemists, engineers, environmental protection groups, car manufacturers, *etc.*

In both cases of role play, intense debate regularly emerged among the students. The students experienced which perspectives contribute to respective decisions. They faced a situation that not only focuses on chemistry content knowledge to provide relevant arguments in the chemistry classroom; arguments stemming from economy, ethics, ecology, and many other fields are needed for a balanced view. The students also learned which arguments are selected by different societal groups and how they are inserted into the debate. The learners even saw that science-related arguments are not introduced by scientists. Representatives of industry and pressure groups purposely select from the available scientific information and use it in a way that supports their interests. The representatives choose from the available arguments and transform them to suit their purposes. Hofstein *et al.* (2011) called this "filtered information". All of these observations were reflected on in the final part of the lesson plan, to contribute to students' skills in understanding and critically reacting to the public debate on sustainability-related socio-scientific questions.

6.4 Teacher Professional Development for Teaching Sustainability in Chemistry Education

Several research studies (*e.g.*, Burmeister *et al.*, 2013) have shown that chemistry teachers are aware of the need for more emphasis on sustainability in the development of contemporary society. There is evidence that both prospective and experienced teachers in at least some countries are generally interested in incorporating societal and sustainability issues into their chemistry teaching. However, teachers often lack the theoretical knowledge of both the content (*e.g.*, the theory and principles of Green Chemistry) and concepts for integrating chemistry learning with ESD (Burmeister and Eilks, 2013a; Burmeister *et al.*, 2013; Ceulemans and Eilks, 2014).

A few years ago, Eilks and coworkers described an initiative to reform chemistry teacher education by incorporating ESD into chemistry teaching in Germany using a combined approach of explorative research, curriculum development, innovation in pre-service teacher education, and teacher continuing professional development. The project centered around an action research project for innovating chemistry teacher education with a focus on ESD (Burmeister and Eilks, 2013b).

The action research approach for this project was derived from a model of participatory action research (PAR) in chemistry curriculum development and teacher continuous professional development suggested by Eilks and Ralle (2002; see Chapter 5 of this book). The project on incorporating ESD into chemistry teacher education (Burmeister and Eilks, 2013b) represents the first approach to transferring the PAR model outlined for school education reform to innovations in teacher education and professional development. (A similar project was later introduced to update chemistry teacher education on the use of information and communication technologies in chemistry teaching based on the same methodology; see Chapter 8 of this book.)

Problem analysis showed many articles and educational policy documents (as discussed in Burmeister *et al.*, 2012) that called for a more thorough implementation of ESD in teacher education programs in general, and chemistry teacher education in particular. The literature suggested that the implementation of learning that addresses both sustainability theories and learning about ESD in chemistry teacher preparation had been insufficient (*e.g.*, Summers and Childs, 2007). This is why the cyclical PAR process was introduced and eventually used for developmental cycles in three consecutive academic years.

Evidence and information about the *a priori* knowledge, attitudes, and beliefs of chemistry student teachers about integrating ESD into chemistry education were hard to come by at the beginning of the process. To overcome the absence of empirical support, explorative analysis of student teacher knowledge of, and attitudes toward ESD were initiated in parallel with the curriculum development. The parallel processes were meant to interact with and influence both curriculum development and the participants' understanding of its effects. One study focused on mapping the knowledge and attitudes of German pre-service chemistry students and postgraduate teacher trainees in the fields of sustainability and ESD where they were connected to chemistry education (Burmeister and Eilks, 2013a).

Burmeister and Eilks (2013a) used questionnaires to map the knowledge and attitudes of student teachers $(N=91)$ at the beginning of their pre-service university training. The questionnaire asked participants questions about modern and theory-based understandings of sustainable development and ESD. A second sample from a compulsory, post-university teacher trainee program $(N=97)$ was also acquired using the same tool. Burmeister and Eilks (2013a) described no major differences between these groups. They found positive attitudes among both student teachers and teacher trainees with regard to strengthening the link between sustainability issues and ESD to secondary-school education. The participants acknowledged that all school subjects should contribute to ESD, but they also acknowledged a specific responsibility for chemistry education. Their overall knowledge about potential topics and pedagogies in this domain was limited, however, and poorly thought out. This work noted a lack of theoretically sound ideas about contemporary concepts of sustainability and ESD in chemistry education. Nevertheless, the study documented positive attitudes toward these topics among the participants. The student teachers and teacher trainees were headed in the right direction when asked for their association with sustainability and ESD. But their ideas were undeveloped and unsupported by theory. Many participants were unable to offer any solid ideas or starting points for integrating ESD into chemistry education. Only a small minority of the participants were able to outline a more or less complete description of what is meant by sustainable development. Almost no one had heard of or could repeat a theoretically-based description of what ESD actually entails.

Most prospective chemistry teachers in the two samples acknowledged that secondary-school education should promote ESD, and that chemistry education should be involved in this effort. Some of them were able to associate topics from the chemistry curriculum with issues of sustainability, such as the question of alternative fuels. Nevertheless, ideas about how teaching might be structured by appropriate pedagogies remained very limited. Both samples explicitly stated that the participants had not been confronted with learning about sustainability or ESD pedagogies during their teacher education. The participants noted that the major sources of their knowledge to that point had been private communication or informational settings such as TV and the Internet. It also became abundantly clear that a theoretical foundation had to start from the very bottom, in terms of both content knowledge and pedagogical content knowledge. The foundation upon which the development of an understanding of these topics could be built needed to include learning about sustainability as such, the role of sustainable thinking in chemistry (*e.g.*, Green Chemistry), and knowledge about practical pedagogies for effectively bringing ESD into the chemistry classroom.

Based on this analysis, a course was created and then refined over three academic years based on observations, reflections, and feedback from the student teachers. The course module lasted 6 weeks, with one 90-minute session per week. Inspired by the exploratory study referenced above (Burmeister and Eilks, 2013a), the coursework started with a self-reflection activity. This activity made the participants explicitly aware of their *a priori* knowledge, their mostly intuitive understanding of sustainability, and their lack of a theoretical foundation for definitions and concepts. The activity made use of the questionnaire from the exploratory study that connected the course participants to the larger sample from the study. This approach allowed the teacher educator to reflect upon and compare participants' thoughts within a larger sample. The analysis showed that many participants might have valuable ideas, but are unaware that such intuitive ideas do not form a comprehensive basis for a theoretically embedded foundation.

Beginning with the exposure of potential deficits in the participants' knowledge, the course then focused on three major areas of learning:

- The historical development and modern concepts of sustainability, in general, and their operationalization in chemistry, in particular, through the concepts of green and sustainable chemistry.

- The basic theories and educational policy legislation concerning ESD with special focus on the practices of chemistry education.
- Adequate pedagogies for acquainting school students with sustainability-based thinking in chemistry classes, promoting their understanding, and increasing their participation skills in societal debates on questions of science and technology.

The approach toward the basic theories behind sustainability was introduced through a short lecture that presented the historical development of the term, the genesis of Agenda 21 (the central document, issued by the UN, suggesting sustainability as a regulatory idea for all political action in the 21st century), and an overview of competing concepts for modeling sustainability. The central focus in this phase was understanding that: (i) modern concepts of sustainability are characterized by different dimensions that are interwoven and contain at least the ecological, economic, and societal dimensions, and (ii) sustainability is always connected to balancing the interests and needs of today's society with the interests and chances of future generations. Learning about the importance of sustainability issues within the context of chemistry and the chemical industry was structured using a WebQuest designed for this course module (Burmeister *et al.*, 2011). The WebQuest (Figure 6.2) introduced problems and issues arising from chemistry and the chemical industry connected to sustainable development. It explained efforts of the chemical industries to contribute to sustainable development, but also presented critical voices. Learning *via* the WebQuest prepared the participants for a role-playing exercise, where both the required effort and represented chances were discussed by different role players. This included discussions of critical roles that question whether the efforts undertaken were carried out in the correct fashion and were sufficiently intense in nature.

Arguments about different ESD theories took place in a jigsaw classroom based on different position papers taken from governmental bodies and educational societies. This cooperative learning scenario was used to analyze and compare position papers from the conference of the German State Ministries of Education (KMK), the German Society for Educational Sciences (DGfE), and the German hub of the UN World Decade of Education for Sustainable Development (Transfer 21). This phase demonstrated the importance placed on ESD by educational theory and educational policy. It also showed that most, if not all, school subjects, including chemistry, should make important contributions to ESD.

The next learning phase for dealing with ESD in school chemistry classrooms was based on a lesson plan that was especially developed by a group of teachers for this purpose (Burmeister and Eilks, 2012). The development of this lesson plan also followed a PAR design (see Chapter 5 of this book), but had innovations in the secondary chemistry classroom as its main goal. The lesson plan dealt with the topic of plastics and covered the basic chemistry and properties of different polymer materials. The lesson

Nachhaltigkeit in der Chemie

Ein WebQuest

| Startseite | Hilfe für Lernende | Hilfe für Lehrkräfte |

Navigation

Einführung

Aufgaben

Ablauf

Quellen

Beurteilung

Abschluss

Worum geht es hier eigentlich?

Ein WebQuest ist eine abenteuerliche Suche im Internet. In diesem WebQuest geht es um Nachhaltigkeit in der Chemie. Du wirst von deinem Lehrer/deiner Lehrerin in eine Gruppe eingeteilt und bekommst eine spezielle Aufgabe zu dem Thema. Du wirst über dieses Portal eine gezielte Recherche durchführen, dazu erhältst du eine Vorauswahl an Links und anderen Quellen. Wichtig ist, dass du mit deiner Gruppe zu einem Ergebnis kommst. Verliere nie den Zweck deiner Recherche aus den Augen!

Viel Spaß.

Die zentrale Aufgabe ist es zu verstehen, was Nachhaltigkeit und nachhaltige Entwicklung bedeuten und die Maßnahmen zur Umsetzung der Nachhaltigkeit in der chemischen Industrie kennenzulernen.

Das WebQuest ist ein Projekt
der Chemiedidaktik der Universität Bremen

Figure 6.2 A WebQuest for pre- and in-service chemistry education on sustainability issues related to chemistry and the chemical industry.

plan focused on ESD by combining the learning of essential chemistry content with information on how to evaluate chemistry products and technologies on the background of sustainability criteria. Within the lesson, the students learned essential chemistry theory, but they also became familiar with the ecological, economic, and societal dimensions of modern sustainability concepts.

The student teachers were asked to mimic consumer test agency workers to experience the interconnectedness of the three sustainability dimensions when evaluating chemistry products and technologies (Burmeister *et al.*, 2014). Within the consumer test agency model, participants were asked to evaluate different sorts of plastics currently addressed by the sustainability debate, all of which have ecological, economic, and societal implications (Figure 6.3). The pupils had to weigh the various dimensions against each other, evaluate the different plastics, and make a final evaluation. They thus learned about the different dimensions, including the fact that different aspects of possible alternatives often negate one another when combined in a comprehensive evaluation. This activity was also mimicked by student teachers during their coursework, then reflected upon with regard to its potential for influencing pupils' learning in a classroom setting. The discussion, using both pro and con arguments about further teaching approaches toward sustainability issues in chemistry education, *e.g.*, an optional board game dealing with Green Chemistry in industrial chemistry (Coffey, 2014), formed another session of the course.

The course closed with a session reflecting on the learned content and present status of ESD implementation in chemistry education. To facilitate this, the four basic models for implementing ESD in chemistry education as described by Burmeister *et al.* (2012) were presented. This phase also referred back to the questionnaire that had been filled out at the beginning of the course. It was now used as a reflection tool on learning progress. The reflections of the participants were also compared with the results of the study by Burmeister and Eilks (2013a) and related to findings from international studies on teachers' and student teachers' understandings and views concerning sustainability issues and ESD in chemistry education. Table 6.1 gives an overview of the different sessions and the pedagogical ideas implemented in the learning process within the course.

The course received favorable and increasing positive feedback from the student teachers (Burmeister and Eilks, 2013b). The positive feedback led to the separation of certain parts of the course content to be used by individual teachers for their continuing professional development, and building on the WebQuest-designed course module (Burmeister *et al.*, 2011) and the lesson plan on bioplastics with the related consumer test activity (Burmeister and Eilks, 2012).

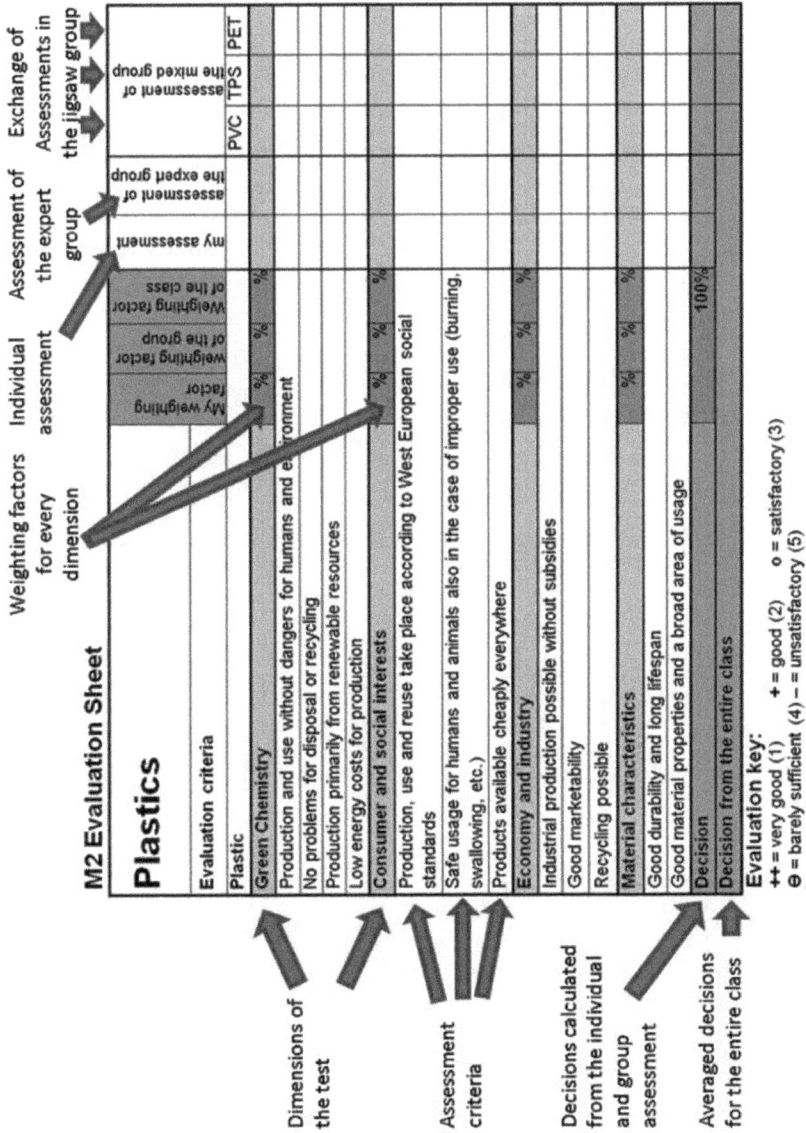

Figure 6.3 Comparing different plastics based on the three dimensions of sustainability.

Table 6.1 Overview of the course module structure [from Burmeister *et al.* (2014), reprinted by permission of the publisher].

Session 1	• Assessing *a priori* knowledge and attitudes toward sustainability and ESD using a research questionnaire • Lecture on the historical genesis and modern concepts of sustainability • Overview on the course and introduction to a WebQuest introducing issues of sustainability, Green Chemistry and the positions of chemical industry and societal pressure groups on sustainability in chemistry-related applications
Session 2	• Working with the WebQuest on issues of sustainability, the concept of Green Chemistry and its perception in society • Role playing a debate between different views of sustainability and Green Chemistry, inspired by the WebQuest
Session 3	Jigsaw classroom on educational policy papers about ESD in secondary-school education
Session 4	Analyzing a lesson plan on teaching about plastics with an ESD focus, which mimics a consumer test to evaluate various sorts of plastics in the foreground of sustainability criteria
Session 5	Further analysis and discussion of teaching materials: • Optional: a board game based on Green Chemistry in the chemical industry • Optional: work on a digital chemistry learning environment on hydraulic fracturing
Session 6	• Lecture presentation summing up the course content • Lecture presentation about basic models of how to connect ESD and chemistry education • Self-assessment of learning success with reference to the initial questionnaire and data on student teachers' knowledge of sustainability and ESD from the accompanying research • Reflection on the course content and structure

6.5 Conclusions

The media is constantly discussing societal and sustainability issues that are related to chemistry (Mandler *et al.*, 2012). The scientifically literate individual is often appalled by the way in which scientific facts and theories are wrongly stated and how the roles of science and technology in both causing and curing these problems are misunderstood. Recently, Hands (2015) completed a study of what happens when the content of a scientific paper evolves through the process of creating a press release that eventually gets reported in a news article. He noted that one of the common elements in this process is the absence of the study's limitations and research methodology in the final article or media presentation. Chemistry education therefore needs to prepare students to deal in an informed manner with science and technology issues as chemistry-literate citizens in a world in which they are able to confront new problems intelligently (Marks *et al.*, 2014), and teacher professional development has to prepare teachers for the challenge of preparing their students for this role in society.

Contextualizing chemistry in societal and sustainability issues can be an effective means of motivating students to deal with chemistry content (Mandler *et al.*, 2012), but also to change the professional view of teachers on how to teach chemistry (Burmeister and Eilks, 2013b). Adding societal, sustainability, and environmental contexts can increase students' perceptions about how relevant chemistry is to real-life problems, which should increase their motivation to learn about it (Eilks, 2002). Motivating the students to study science, while, for example, providing a good understanding of chemical concepts, is ranked among the most important concerns in chemistry education (Zoller and Pushkin, 2007).

The inclusion of societal, sustainability, and environmental contexts in chemistry education can have a significant impact on students' awareness of how chemistry is connected to the real world (Eilks, 2002). Students become more aware of the relationships between chemistry and the society in which they live (Marks and Eilks, 2009). Basic chemistry knowledge is crucial for promoting sustainable development and improving people's capacity to address environmental and developmental issues, and thus needs to be incorporated as an essential part of chemistry pre- and in-service professional development. Integrating chemistry learning with sustainability and environmental contexts is very much in line with the idea of ESD. With respect to a critical approach, this represents a context-based or SSI-based mode of integrating ESD into chemistry learning as suggested by Burmeister *et al.* (2012).

However, learning chemistry by focusing on sustainability in an SSI mode should have a broader impact. Learning chemistry can directly help citizens to better react to sustainability and environmental issues by providing data and theories that define the problem, assess its seriousness, and explain its causes. The environmental literacy report by the National Environmental Education and Training Foundation in the USA recommends that people receive concrete suggestions on how to change their behavior, with an emphasis on how they can join others who are doing the same (Coyle, 2005). The report noted a positive correlation between environmental knowledge and environmental behaviors: the more you know, the more likely you are to take active positive steps toward improving the environment (Coyle, 2005).

Acting for sustainable development encompasses participation in societal discourse and decision making. The discussion in this chapter makes clear that increasing learning success in pure chemistry knowledge is not sufficient to understand or to participate in the societal debate on SSI concerning questions of chemistry and related chemical technology (Burmeister *et al.*, 2012). A broad skill set is needed to act responsibly in the societal context. Communication and decision-making skills need to be developed, as well as skills in analyzing how science- and chemistry-related arguments are used in society (Eilks *et al.*, 2014), and chemistry education should contribute to developing these skills. Integrating chemistry learning with issues of sustainable development offers great opportunities to create learning environments that contribute general skill development beyond pure chemistry. The examples presented in this chapter show that corresponding lesson plans have great potential to both increase motivation and promote chemistry learning, as well as help in the development of general educational skills.

6.6 Summary

- Contemporary chemistry education should address all learners and promote their skills for a self-determined life and participation in society. To fulfill this aim, curricula are needed that focus on SSI and the linkages among chemistry, technology, society, and the environment.
- ESD is a highly relevant educational goal. With the importance of chemistry for the sustainable development of our future, chemistry education has a special responsibility to contribute to ESD. This can be done by, for example, dealing with issues of modern materials, alternative energy supplies, or minimizing resource consumption in the chemistry classroom, from both a chemical and societal perspective.
- Societal issues related to chemistry and the question of how to achieve sustainable development of our future are ever-changing challenges. The constant change in topics and issues related to societal issues and questions of sustainable development demands continuing efforts in the professional development of all chemistry teachers. Corresponding professional development needs to address both the related chemical content, *e.g.*, recent developments in nanochemistry or the synthesis of new materials, and corresponding curricula with their related pedagogies.

References

Burmeister M. and Eilks I., (2012), An example of learning about plastics and their evaluation as a contribution to Education for Sustainable Development in secondary school chemistry teaching, *Chem. Educ. Res. Pract.*, **13**, 93–102.

Burmeister M. and Eilks I., (2013a), German chemistry student teachers' and trainee teachers' understanding of sustainability and education for sustainable development, *Sci. Educ. Int.*, **24**, 167–194.

Burmeister M. and Eilks I., (2013b), Using Participatory Action Research to develop a course module on Education for Sustainable Development in pre-service chemistry teacher education, *Centre Educ. Pol. Stud. J.*, **3**, 59–78.

Burmeister M., Jokmin S. and Eilks I., (2011), Bildung für nachhaltige Entwicklung und Green Chemistry im Chemieunterricht [Education for sustainable development and green chemistry in chemistry teaching], *Chem. kon.*, **18**, 123–128.

Burmeister M., Rauch F. and Eilks I., (2012), Education for Sustainable Development (ESD) and secondary chemistry education, *Chem. Educ. Res. Pract.*, **13**, 59–68.

Burmeister M., Schmidt-Jacob S. and Eilks I., (2013), German chemistry teachers' knowledge and PCK of Green Chemistry and education for sustainable development, *Chem. Educ. Res. Pract.*, **14**, 169–176.

Burmeister M., von Döhlen J. and Eilks I., (2014), Learning about the different dimensions of sustainability by applying the product test method in science classes, in Thomas K. D. and Muga H. E. (ed.), *Handbook of Pedagogical Innovations for Sustainable Development*, Hershey: IGI Global, pp. 154–169.

Bybee R. W. and Loucks-Horsley S., (2000), Supporting change through professional development, in Resh B. (ed.), *Making Sense of Integrated Science: A Guide for High Schools*, Colorado Springs: BSCS, pp. 41–48.

Ceulemans G. and Eilks I., (2014), Will Flemish chemistry teachers soon implement ESD? in Eilks I., Markic S. and Ralle B. (ed.), *Science Education Research and Education for Sustainable Development*, Aachen: Shaker, pp. 231–236.

Coffey M., (2014), Green Chemistry: classroom implementation of an educational board-game illustrating environmental sustainability in chemical manufacturing, in Thomas K. D. and Muga H. E. (ed.), *Handbook of Research on Pedagogical Innovations for Sustainable Development*, Hershey: IGI Global, pp. 453–473.

Coyle K., (2005), *Environmental Literacy in America: What Ten Years of NEETF/ Roper Research and Related Studies Say about Environmental Literacy in the US*, Washington: NEETF.

Eilks I., (2002), Teaching 'Biodiesel': a sociocritical and problem-oriented approach to chemistry teaching, and students' first views on it, *Chem. Educ. Res. Pract.*, **3**, 67–75.

Eilks I. and Hofstein A., (2014), Combining the question of the relevance of science education with the idea of education for sustainable development, in Eilks I., Markic S. and Ralle B. (ed.), *Science Education Research and Education for Sustainable Development*, Aachen: Shaker, pp. 3–14.

Eilks I., Nielsen J. A. and Hofstein A. (2014), Learning about the role of science in public debate as an essential component of scientific literacy, in Bruguière C., Tiberghien A., Clément P. (eds.), *Topics and Trends in Current Science Education*, Dordrecht: Springer, pp. 85–100.

Eilks I. and Ralle B., (2002), Participatory Action Research in chemical education, in Ralle B. and Eilks I. (ed.), *Research in Chemical Education – What Does This Mean?* Aachen: Shaker, pp. 87–98.

Eilks I., Ralle B., Rauch F. and Hofstein A., (2013), How to balance the chemistry curriculum between science and society, in Eilks I. and Hofstein A. (ed.), *Teaching Chemistry – A Studybook*, Rotterdam: Sense, pp. 1–36.

Ezra L., Skolnick B. and Aghbariya G., (2012), Can used oil be the next generation fuel? unpublished module developed in the framework of the PROFILES Project funded by the European Community's 7th Framework Program.

Feierabend T. and Eilks I., (2011), Teaching the societal dimension of chemistry along a socio-critical and problem-oriented lesson plan on the use of bioethanol, *J. Chem. Educ.*, **88**, 1250–1256.

Fensham P. J., (1985), Science for all: a reflective essay, *J. Curr. Stud.*, **17**, 415–435.

Hands Jr. M. D., (2015), Public understanding of chemistry research in print news, Ph.D. Dissertation, Purdue University, West Lafayette, USA.

Haney J. J., Lumpe A. T., Czerniak C. M. and Egan V., (2002), From beliefs to actions: the beliefs and actions of teachers implementing change, *J. Sci. Teach. Educ.*, **13**(3), 171–187.

Harms N. and Yager R. E., (1981), *What Research Says to the Science Teacher*, Washington: NSTA.

Hofstein A., Eilks I. and Bybee R., (2011), Societal issues and their importance for contemporary science education: a pedagogical justification and the state of the art in Israel, Germany and the USA, *Int. J. Sci. Math. Educ.*, **9**, 1459–1483.

Hofstein A., Mamlok R. and Carmeli M., (1997), Science teachers as curriculum developers of Science for All, *Sci. Educ. Int.*, **8**(1), 26–29.

Joyce B. and Showers B., (1983), *Power and Staff Development through Research on Training*, Alexandria: Association for Supervision and Curriculum Development.

Loucks-Horsley S. and Matsumoto C., (1999), Research on professional development for teachers of mathematics and science: the state of the scene, *Sch. Sci. Math.*, **99**, 258–271.

Mamlok-Naaman R., Katchevich D., Yayon M., Burmeister M. and Eilks I., (2015), Learning about sustainable development in socio-scientific issues-based chemistry lessons on fuels and bioplastics, in Zuin V. G. and Mammino L. (ed.), *Worldwide Trends in Green Chemistry Education*, Cambridge: RSC, pp. 45–60.

Mandler D., Mamlok-Naaman R., Blonder R., Yayon M. and Hofstein A., (2012), High-school chemistry teaching through environmentally oriented curricula, *Chem. Educ. Res. Pract.*, **13**, 80–92.

Marks R. and Eilks I., (2009), Promoting Scientific Literacy using a socio-critical and problem-oriented approach in chemistry education: concept, examples, experiences, *Int. J. Environ. Sci. Educ.*, **4**, 131–145.

Marks R., Stuckey M., Belova N. and Eilks I., (2014), The societal dimension in German science education – from tradition towards selected cases and recent developments, *EurAsia J. Math. Sci. Technol. Educ.*, **10**, 285–296.

Matlin S. A., Mehta G., Hopf H. and Krief A., (2015), The role of chemistry in inventing a sustainable future, *Nat. Chem.*, 7, 941–943.

Mitchener C. P. and Anderson R. D., (1987), Teachers' perspective: developing and implementing an STS curriculum, *J. Res. Sci. Teach.*, **26**, 351–369.

Penick J., (1984), *Focus on Excellence: Science/Technology/Society*, Washington: NSTA.

Stolz M., Witteck T., Marks R. and Eilks I., (2013), Reflecting socio-scientific issues for science education coming from the case of curriculum development on doping in chemistry education, *EurAsia J. Math. Sci. Technol. Educ.*, **9**(4), 273–282.

Stuckey M., Hofstein A., Mamlok-Naaman R. and Eilks I., (2013), The meaning of 'relevance' in science education and its implications for the science curriculum, *Stud. Sci. Educ.*, **49**, 1–34.

Summers M. and Childs A., (2007), Student science teachers' conceptions of sustainable development: an empirical study of three postgraduate training cohorts, *Res. Sci. Technol. Educ.*, **25**, 307–327.

Sweeney A. E., (2001), Incorporating multicultural and science/technology/ society issues into science teacher education courses: successes, challenges and possibilities, *J. Sci. Teach. Educ.*, **12**, 1–28.

Tobin K., Tippins D. J. and Gallard A. J., (1994), Research on instructional strategies for teaching science, in Gabel D. L. (ed.), *Handbook of Research on Science Teaching and Learning*, New York: Macmillan, pp. 45–93.

UNCED, (1992), Agenda 21, retrieved from http://www.un.org/esa/dsd/ agenda21/.

Wiggins G. and McTighe J., (1998), *Understanding by Design*, Upper Saddle River: Prentice Hall.

WCED, (1987), *Our Common Future*, Oxford: Oxford University Press.

Yager R. E., (1996), *Science/Technology/Society as Reform in Science Education*, Albany: SUNY Press.

Zoller U. and Pushkin D., (2007), Matching Higher-Order Cognitive Skills (HOCS) promotion goals with problem-based laboratory practice in a freshman organic chemistry course, *Chem. Educ. Res. Pract.*, **8**, 153–171.

CHAPTER 7

Professional Development of Chemistry Teachers to Teach Effectively in the Chemistry Laboratory

Laboratory activities have long had a distinctive and central role in the science curriculum as a means of making sense of the natural world. The laboratory in science education and in general, and in chemistry education in particular, is a unique learning environment. Thus, it deserves unique assessment methods and a unique model to prepare new chemistry teachers in the pre-service mode of their career and as part of continuous professional development (CPD) in the in-service professional mode. In this chapter we discuss the uniqueness of the chemistry laboratory and the special skills chemistry teachers need in order to teach effectively, to develop through inquiry-type high-order learning and thinking skills, and develop interest in and motivation for learning chemistry among students.

7.1 The Chemistry Teacher's Behavior in the Chemistry Laboratory

Since the 19th century, when schools began to teach science systematically, the laboratory has become a distinctive feature of science education. After the First World War, and with the rapid increase in scientific knowledge, the laboratory was used mainly as a means of confirming and illustrating information previously learned in a lecture or from textbooks. With the reform in science education in the 1960s in many countries, the ideal

Advances in Chemistry Education Series No. 1
Professional Development of Chemistry Teachers: Theory and Practice
By Rachel Mamlok-Naaman, Ingo Eilks, George Bodner and Avi Hofstein
© Rachel Mamlok-Naaman, Ingo Eilks, George Bodner and Avi Hofstein 2022
Published by the Royal Society of Chemistry, www.rsc.org

became to engage students with investigations, discoveries, inquiry, and problem-solving activities. In other words, the laboratory became the core of the science learning process (Tamir, 1989).

The National Science Education Standards (NRC, 1996, p. 23) defines such learning activities (*e.g.*, inquiry) as: "The diverse ways in which scientists study the natural world and propose explanations based on the evidence derived from their work. Scientific inquiry also refers to the activities through which students develop knowledge and understanding of scientific ideas, as well as an understanding of how scientists study the natural world."

For years, science educators have suggested that many benefits accrue from engaging students in science laboratory activities (Tobin, 1990; Hofstein and Lunetta, 2004). For example, Tobin (1990) wrote that: "Laboratory activities appeal as a way of allowing students to learn with understanding and at the same time engage in the process of constructing knowledge by doing science." In curricular-type projects developed in the 1960s, the laboratory was intended to be a place for inquiring about, developing, and testing theories, as well as providing students with the opportunity to "practice being a scientist". Many research studies (summarized, for example, by Bates, 1978; Hofstein and Lunetta, 1982) were conducted with the goal of exploring the effectiveness of the laboratory for attaining the many objectives (both cognitive and affective) suggested in the science education literature. The traditional list of objectives included:

- Understanding of scientific concepts
- Interest and motivation
- Attitude toward science
- Scientific practical skills and problem-solving abilities
- Scientific habits of mind
- Understanding the nature of science (NOS)
- The opportunity to *do* science

Over the years, hundreds of papers and essays have been published with the aim of exploring and investigating the uniqueness of the science laboratory in general and its educational effectiveness in particular. In addition, it was widely believed that the laboratory provides the only place in school where certain kinds of skills, abilities, and understanding can be developed (Lazarowitz and Tamir, 1994). However, precisely what kinds of objectives and aims are attained in the laboratory depends on a wide range of factors. We suggest that, among others, these will include the teacher's goals; expectations, subject and pedagogical content knowledge, as well as the degree of relevance to the topic, the students' abilities and interests, and many other logistical and economic considerations related to the school setting and facilities (see Table 7.1).

It should be noted that some of these goals, such as "to enhance understanding of scientific ideas", coincide with the broad goals for science

Table 7.1   Suggested goals for laboratory activity (Bennet, 2006, pp. 78–79).

Goals for laboratory activity
To encourage accurate observation and description
To make scientific phenomena more real
To enhance understanding of scientific ideas
To arouse and maintain interest (particularly with younger pupils)
To promote a scientific method of thought

education that are not necessarily laboratory-based. The teacher should be in a position to judge whether the laboratory is the most effective learning environment for attaining a certain objective while teaching a particular topic. Teachers should be aware that there has been a great deal of discussion and numerous research studies on which goals are in fact better achieved through laboratory instruction than through other instructional (pedagogical) approaches (Hofstein and Lunetta, 1982, 2004). The many research studies and essays that were cited in these reviews criticized the tradition of conducting experiments without a clear purpose or goal. In addition, they revealed a significant mismatch between teachers' goals for learning in the science laboratory and those that were originally defined by curriculum developers and the science education milieu.

7.2   Research-based Ideas Related to Learning in the Science Laboratory

The main goal of this chapter is to argue and demonstrate that since the laboratory in science education is a unique learning environment (Lazarowitz and Tamir, 1994; Lunetta *et al.*, 2007), it warrants unique professional development models to train teachers to teach effectively in chemistry laboratories.

The distinct and central role of laboratory activities in the science curriculum has led science educators to suggest the benefits of engaging students in these activities (Hofstein and Lunetta, 1982, 2004; Tobin, 1990; Lazarowitz and Tamir, 1994; Lunetta, 1998; Lunetta *et al.*, 2007). More specifically, they have suggested that, when properly developed, inquiry-centered laboratories have the potential to enhance students' meaningful learning, conceptual understanding, and understanding of the NOS. This suggested that inquiry-type laboratories are central to learning science, because students are involved in the process of conceiving problems and scientific questions, formulating hypotheses, designing experiments, gathering and analyzing data, and drawing conclusions about scientific problems or phenomena.

Today we are entering a new era of reform in science education. Both the content and pedagogy of science learning and teaching are being scrutinized (Bybee, 2000; NRC, 1996). One of the general characteristics of the current reform is the articulated change in goals for science teaching and learning, namely, that science education should be targeted to all students (attaining scientific literacy for all students) and should be extended beyond the preparation of science-oriented students for academic careers in the sciences. This is, in fact, also a call for rethinking the goals for learning in the laboratory and from laboratory work. There are several buzzwords that characterize the current reform. Among them are student-centered learning, learning by inquiry, and development of high-order learning skills such as argumentation, metacognition, and the asking of relevant (to the experimental situation) questions. Inquiry in the context of science learning in general and in the science laboratory in particular is one of the most important components of this reform (Lunetta, 1998; Bybee, 2000; Sere, 2002; Hofstein and Lunetta, 2004). Bybee (2000) suggested that inquiry in terms of skills and abilities includes the following components: identifying and posing scientifically oriented questions, forming hypotheses, designing and conducting scientific investigations, formulating and revising scientific explanations, and communicating and defending scientific arguments. It is suggested that many of these abilities and skills are in alignment with those that characterize inquiry-type laboratory work (practical work that includes project-based learning), an activity that places the student at the center of the learning process (see also Hofstein *et al.*, 2004, 2005). Tobin (1990) based his review on constructivist ideas of providing students with experiences that would enable meaningful learning in the science laboratory. He wrote that "laboratory activities appeal as a way of allowing students to learn with understanding and, at the same time, engage in the process of constructing knowledge by doing science" (p. 405). To attain this goal, he suggested that in the laboratory, students be provided with opportunities to reflect on findings, clarify understandings and misunderstandings with peers, and consult a range of resources including teachers, books, websites, and other learning materials. His review reported that such opportunities rarely exist because teachers in the laboratory are so often preoccupied with technical and managerial activities. Similarly, Hodson (1993) suggested that although teachers generally profess belief in the value of student-driven, open, practical investigation, in general their teaching practices in the laboratory fail to support this claim. He also argued that the research literature fails to provide evidence of standard school laboratory activities encouraging knowledge construction. He was critical of the research literature:

> Despite the very obvious differences among, for example, practical exercises designed to develop manipulative skills or to measure 'physical constraints', demonstration-type experiments to illustrate certain key-concepts, and inquiries that enable children to conduct their own investigations, there is a tendency for researchers to lump them all together under the same umbrella title of practical work. (p. 97)

Tobin wrote that teachers' interpretation of practical activities should be elaborated upon and made a part of the research design because a laboratory session could consist of open-ended inquiry in one classroom and be more didactic and confirmatory in another teacher's classroom.

Based on their review of the literature regarding the use of laboratory activities, Lazarowitz and Tamir (1994) joined the long list of authors claiming that the potential of the laboratory as a medium for teaching and learning science is enormous. They wrote that the laboratory is the *only* place in school where certain kinds of skills and understanding can be developed. They suggested, in agreement with others, that one of the complicating factors associated with research on the effectiveness of the school laboratory is that often, the goals articulated for learning in the laboratory refer to experiences in the laboratory, and related assessment practices have remained relatively unchanged. This is almost synonymous with suggestions articulated for learning science more generally. Hurd (1983) claimed that much of the practical work is purposeless and often, the explicit objectives of the practical work do not coincide with the purpose of the practical experiences. They also claimed that many practical tasks have too many different teaching/learning objectives to focus on during instruction. Similarly, Sere (2002), reporting on a long-term project (Lab-Work in Science Education) conducted in seven European countries wrote that:

> The intention [of the study] was to address the problem of the effectiveness of lab-work, which in most countries is recognized as being essential to experimental sciences, but which turns out to be expensive and less effective than wished. (p. 624)

The project focused mainly on the effectiveness of laboratory work conducted in the context of science learning in upper secondary schools. Information on practice was gathered through 23 case studies, surveys, and a tool that maps and describes the domain of laboratory work. Sere (2002) reported that the objectives typically articulated for laboratory work (*i.e.*, understanding theories, concepts, and laws; conducting experiments; learning processes and approaches; applying knowledge to new situations) were too numerous and comprehensive for teachers to address successfully in individual laboratory sessions. As a consequence, she suggested that the scope of the objectives for specific laboratory activities should be limited. Science curriculum developers and science teachers should make conscious choices among specific learning objectives for specific laboratory activities and clearly articulate the specific objectives for their students. Sere's (2002) "targeted lab-work" project produced a series of recommendations, including the need for each laboratory activity to be supported by a particular strategy organized within a coherent long-term program plan with varied types of laboratory work.

Gunstone and Champagne (1990) wrote that helping students develop scientific ideas from practical experiences is a very complex process and that students generally do not have sufficient time or encouragement to express their interpretations and beliefs and to reflect on central ideas in the laboratory. Research on learning in the school laboratory makes clear that to understand their laboratory experiences, students must manipulate *ideas* as well as materials in the school laboratory and they must be helped to contrast (and align) their findings and ideas with the concepts of the contemporary scientific community. Manipulating materials in the laboratory is *not* sufficient for learning contemporary scientific concepts. This accounts for the failure of "cookbook" laboratory activities and relatively "unguided" discovery activities to promote the desired scientific understanding. Several studies suggested that while laboratory investigations offer excellent settings in which students can make sense of phenomena and in which teachers can better understand their students' thinking, laboratory inquiry alone is *not* sufficient to enable students to construct the complex conceptual understandings of the contemporary scientific community (Lunetta, 1998). In the laboratory, students should be encouraged to articulate and share their ideas, in order to help them perceive discrepancies among their ideas, those of their classmates, and those of the scientific community.

At the end of the 20th century, an understanding emerged from cognitive sciences that learning is contextualized and that learners construct knowledge by solving genuine, meaningful problems. The school science laboratory can offer students some control of their activities, thereby enhancing their perception of sense of *ownership* and *motivation* (Johnstone and Al-Shuaili, 2001). This environment can be particularly well suited to providing a meaningful context for learning, determining and challenging students' deeply held ideas about natural phenomena, and constructing and reconstructing their ideas. Although a complex process, meaningful learning in the laboratory *can* occur if students are given sufficient time and opportunities to interact, reflect, explain, and modify their ideas. Engaging in *metacognitive* behaviors of this kind enables students to elaborate and apply their ideas; the process can promote conceptual understanding as well as the development of problem-solving skills. The challenge is to help learners take control of their own learning in the search for understanding, while providing opportunities that encourage them to ask questions, suggest hypotheses, and design investigations, "minds-on as well as hands-on" (Gunstone and Champagne, 1990).

7.3 New Era – New Goals: Learning in and from the Science Laboratory in the 21st Century

We are operating in an era in which high-order learning skills are considered to be as important as the content of science (Hofstein and Kind, 2012). High-order thinking/learning skills and activities in the context of learning science are considered to be complex and non-algorithmic, and to involve applications of multiple criteria instead of memorizing facts (Livingstone, 1997). These activities include asking research questions, solving authentic problems, argumentation, metacognitive skills, drawing conclusions, making comparisons, dealing with controversies, and taking a stand. Gunstone and Champagne (1990) claimed that meaningful learning in the laboratory occurs when students are given ample opportunities for inter-action and reflection in order to initiate discussion. It is suggested that some of these skills could be developed as part of inquiry-based science labora-tories. Many of these abilities and skills are in alignment with those that characterize inquiry-based chemistry laboratory work, an activity that places the student at the center of the learning (Sandoval, 2005). Researchers claim that learning in the laboratory might provide a constructivist environment that fosters high-order thinking, and metacognitive and argumentative skills (Kind, 2003). Here we elaborate on two of these variables, namely the development of argumentative skills and the ability to ask high-level and relevant questions, in the context of the science laboratory.

Based on their research, Kipnis and Hofstein (2008) linked metacognitive skills to various stages of inquiry-oriented experiments: (i) while asking questions and choosing an inquiry question, the students revealed their thoughts about their and their partners' suggested questions. At this stage, *metacognitive declarative knowledge* is expressed; (ii) while choosing the inquiry question, the students expressed their *metacognitive procedural knowledge* by choosing the one that leads to conclusions; (iii) while performing their own experiment and planning changes and improvements, the students demonstrated the *planning* component of *regulation of cognition*; (iv) at the final stage of the inquiry activity, when the students write their report and have to draw conclusions, they utilized *metacognitive conditional knowledge*; (v) throughout the activity, the students made use of the *monitoring* and *evaluating* components related to *regulation of cognition*. In this way, they examined the results of their observations to decide whether they are logical.

7.3.1 Scientific Argumentation and Epistemologies

When Driver *et al.* (2000) presented their introduction to argumentation in science education they quickly indicated the relevance of practical work. They saw argumentation as correcting the misinterpretation of the scientific method that had dominated much of science teaching in general and practical work in particular. Rather than focusing on the stepwise series of actions carried out by scientists in experiments, they claimed that focus should be directed toward the *epistemic practice* involved when developing and evaluating scientific knowledge. We sense two overlapping learning aims. First, that students should understand the scientific standards and their guiding epistemologies, and second, that they should be able to apply these standards in their own argumentation.

We find many ways of approaching research into students' epistemological understanding and argumentation skills. One contribution comes from psychologists who identify scientific argumentation as the key element of scientific thinking (Kuhn, 1988). Those authors worked from the perspective that certain reasoning skills related to argumentation are domain-general. People who are good at scientific argumentation are able to (i) think *about* a scientific theory, rather than just think *with* it; (ii) encode and think about evidence in a similar way, and in this way distance evidence from theory; (iii) put aside their personal opinions about what is "right" and instead weigh the theoretical claim against the evidence. Kuhn *et al.* (2000) demonstrated how these abilities develop naturally from childhood to adulthood, but also that their quality varies among individuals. Scientists are good at this thinking because it is embedded in their culture and, most importantly, explicit training in the science laboratory seems to help (Kuhn *et al.*, 2000).

Several research studies have indicated that the development of students' argumentation skills and science epistemologies is rather complex. Students, for example, may hold some beliefs about professional science and very different beliefs about their own practices with inquiry at school; in other words, they may have one set of *formal* epistemologies and another set of *personal* epistemologies (Sandoval, 2005). Many years of teaching "ideas and evidence" in the UK through practical investigations illustrates this complexity (Solomon, 1994; Driver *et al.*, 1996). The overall picture has been that students become good at doing specific types of routine experiments, and solving these using school-based strategies rather than a general understanding of formal scientific epistemologies (Kind, 2003). Many studies have also related the problems with developing epistemological views and practices in school science to the teachers' background and competencies. A connection was found between the epistemological beliefs expressed by teachers and their preferred ways of teaching (Hashweh, 1996), but the relationship was not a straightforward one: teachers with naïve epistemological beliefs most easily supported teaching "real science" in the school laboratory. For teachers with a more sophisticated epistemological understanding of science, the relationship was more complicated. They tended to disconnect "real science" from "school science" and more rarely allowed their epistemological beliefs to be reflected in their teaching practice (Barnett and Hodson, 2001). They also seemed to separate science from students, by treating them more as "spectators" of science.

A way forward to understanding how practical work may contribute to the development of students' epistemological understanding and argumentation skills may be to look more closely at the "teaching ecology" of the laboratory (Jimenez-Aleixandre *et al.*, 2000). It is strongly believed that bringing argumentation into science classrooms requires the enactment of contexts that transform them into knowledge-producing communities, encouraging dialogic discourse and various forms of cognitive, social, and cultural interactions among learners (Duschl and Osborne, 2002). The ecology that invites this practice is created through the social and physical environment of the laboratory tasks (Chinn and Malhotra, 2002) and the organizational principles used by the teacher (Scott, 1998) A reconsideration of all of these factors is therefore needed for the science laboratory to contribute meaningfully and effectively to the new learning goals.

7.3.2   Argumentation in the Science Laboratory

Several researchers who focused on the issue of argumentation suggested that the inquiry-type laboratory in science education can provide opportunities for students to develop argumentation skills (see a detailed discussion in Hofstein and Kind, 2012). However, only a few studies have been conducted with the goal of accepting or rejecting this assumption. For example, Rickey and Stacy (2000) found that students who participated in guided inquiry-type laboratories were better at evaluating evidence obtained from their research.

Two recent reports looked at the nature of experiments as a platform for evoking argumentation both quantitatively (number of arguments) and qualitatively (level of arguments). Kind *et al.* (2011) investigated the quality of argumentation among 12- to 13-year-old students in the UK in the context of a secondary school physical science program. Their study explored the development of argumentation in students who undertook three different designs of laboratory-based tasks. The tasks described in their paper involved the students in the following: collecting and making sense of data, collecting data for addressing conflicting hypotheses, and paper-based discussions on the pre-collected data phase of an experiment. Their findings showed that the paper-based task generated a larger number of arguments per unit time than the other two tasks. In addition, they found that to encourage the development of high-level and authentic argumentation, the practice that generally exists in science laboratories in England needed to be changed. They suggested that more rigorous and longitudinal research is needed to explore the potential of the science laboratory as a platform for the development of students' ability to argue effectively and in an articulate way.

The second study was conducted in Israel in the context of 12 years of research and development of inquiry-type laboratories in the context of upper secondary school, grades 10–12 (for more details about the philosophy and rationale of the project, see Hofstein *et al.*, 2004). The implementation and effectiveness of this project was researched intensively and comprehensively and reported in series of manuscripts (Hofstein *et al.*, 2004; Kipnis and Hofstein, 2008). The research study conducted by Katchevich *et al.* (2013) focused on the process by which students construct arguments in the chemistry laboratory while conducting different types of experiments. It was found that *inquiry-type* experiments can potentially serve as an effective platform for formulating arguments, owing to the special features of this learning environment. The discourse conducted during inquiry-type experiments was found to be rich in arguments, whereas during *confirmatory-type* experiments, arguments were sparse. In addition, it was found that the arguments, which were developed during the discourse of an inquiry-type experiment, were generated during the following stages of the inquiry process: hypothesis-building analysis of the results, and drawing appropriate conclusions. On the other hand, the confirmatory-type experiment revealed a small number of arguments which were of a low level.

Based on detailed analysis of the discourse in the chemistry laboratory, we may conclude that the open-ended inquiry experiments stimulate and encourage the construction of arguments, especially at the stages of hypothesis definition, results analysis, and conclusion drawing. Some arguments were raised by individuals, and some by the group. Both types of arguments consisted of explanations and scientific evidence, which linked the claims to the evidence. Therefore, it is suggested that the learning environment of open-ended inquiry experiments is a platform for raising arguments. In this study, we wanted to point out the main factors that stimulate raising arguments in open-ended inquiry experiments, as well as to characterize situations in which argumentation develops into a significant discourse.

The key question relevant to the current chapter is: "How are laboratories used and what is the chemistry teacher's key role in this rather important and unique learning mode?" Based on the research literature (Lazarowitz and Tamir, 1994), it is well established that the chemistry laboratory provides a unique mode of teaching, learning, and assessment. Thus, the teacher's role in this unique learning environment should be to present different behaviors and tasks. Several studies have reported that very often, teachers involve students principally in relatively low-level, routine activities in laboratories and that teacher–student interactions focus principally on low-level procedural questions and answers. Marx *et al.* (1998) reported that science teachers often have difficulty helping students ask thoughtful questions, design investigations, and draw conclusions from the data. Similar findings were reported regarding the chemistry laboratory setting (De Carlo and Rubba, 1994). More recently, Abrahams and Millar (2008) investigated the effectiveness of practical work by analyzing a sample of 25 "typical" science lessons involving practical work in English secondary schools. They concluded that the teachers' focus in these lessons was predominantly on making students manipulate physical objects and equipment. Hardly any of the teachers focused on the cognitive challenge of linking observations and experiences to conceptual ideas. Nor was there any focus on developing students' understanding of scientific inquiry procedures. These findings echo the situation at any given time in the history of school science. Basic elements of teachers' implementation of practical work seem not to have changed over the last century; students still carry out recipe-type activities that are supposed to reflect science procedures and teach science knowledge, but which in general fail at both. This is not to say that everything is the same. Science education has moved forward in the last few decades and improved teachers' professional knowledge and classroom practice, but this improvement has not caught up sufficiently with the challenges of using laboratory work in an efficient and appropriate way. Teachers still do not perceive of what is required to make laboratory activities serve as a principal means of enabling students to construct meaningful knowledge of science, and they do not engage students in laboratory activities in ways that are likely to promote the development of science concepts. In addition, many teachers do not believe that helping students understand

how scientific knowledge is developed and used in a scientific community is
an especially important goal of laboratory activities for their students.

Today's conclusion has therefore not changed substantially from that of
Woolnough and Alsop (1985), who claimed that:

> Teachers at present are ill prepared to teach effectively in the laboratory.
> A major reason is that most science teachers were themselves brought-
> upon a diet of content dominated cookery book-type practical work and
> many have got in their habit of propagating it themselves. (p. 80)

Aligned with this situation for teachers, we find a matching picture for
students' experiences with laboratory teaching material. In the USA, Domin
(1999), in the context of learning chemistry in colleges, found that students
are seldom given the opportunity to use high-level cognitive skills or to
discuss substantive scientific knowledge associated with the investigation,
and many of the tasks presented to them continue to follow a 'cookbook'
approach, concentrating on the development of lower-level skills and abilities. The reviews discussed earlier in this chapter reported a mismatch
between the goals articulated for the school science laboratory and what
students generally do there. Ensuring that students' experiences in the laboratory are aligned with the stated goals for learning demands that teachers
explicitly link decisions regarding laboratory topics, activities, materials, and
teaching strategies to the desired outcomes for student learning. The body of
published research suggests that far more attention must be paid to the
crucial roles of the teacher and other sources of guidance during laboratory
activities, and researchers must also be diligent in examining the many
variables that interact to influence the learning that occurs in the complex
classroom laboratory.

It has been suggested that inquiry-centered laboratories have the potential
to enhance students' meaningful learning, conceptual understanding,
and their understanding of the NOS. Inquiry-type experiences in the science
laboratory should be conducted in the context of, and integrated with, the
concept being taught.

The *National Science Education Standards* (NRC, 1996) reaffirm the
conviction that inquiry is central to the achievement of scientific literacy.

Teaching and learning science by inquiry is an approach that places
the student at the center of the learning process. Such an approach poses
challenges to both students and their teachers (Johnstone and Wham,
1982). Education has moved forward in the last few decades and improved
teachers' professional knowledge and classroom practice, but this improvement has not sufficiently caught up with the challenges of using
laboratory work in an efficient and appropriate way. The biggest challenge
for practical work, historically and today, is to change the practice of
"manipulating equipment not ideas". The typical laboratory experience in
school science is a "hands-on" but not a "minds-on" activity. This problem
is related to teachers' fear of losing control in the classroom by giving

students more responsibility for their learning. Also to blame for the current situation are an assessment practice that does not pay enough attention to high-order thinking, and a long tradition of developing foolproof laboratory tasks that guide students through activities without requiring any deeper reflection. The review in this chapter demonstrates a relationship between these problems in practical work and the "common sense" ideas about science inquiry as a stepwise method.

It has taken science education research a long time to reveal this practice, analyze its underlying rationales and to present alternatives. The development has required a move away from quantitative research methods, which were not sensitive to students' learning in the laboratory, toward more authentic ways of studying what actually goes on in the laboratory. It has also required a thorough analysis of the NOS inquiry and what makes someone good at it. The alternatives that are prominent today combine sociocultural perspectives on science and learning, but also link to new aims for school science as an important provider of skills and knowledge for citizenship.

Today, we may claim that science education is in a better position than ever before for developing a meaningful and appropriate practice for laboratory work. The situation is most promising because of the results and knowledge that have been accumulated and achieved. There are many places to start for new development of laboratory teaching strategies and professional development of teachers. These and other tasks call for science education researchers to continue engaging with practical work and help develop this area further.

7.4   Teaching in an Inquiry-type Laboratory

To teach science using the inquiry method, teachers need to undergo intensive and comprehensive professional development that will equip them with the knowledge and skills needed to counsel their students through inquiry-type experiences (Windschitl, 2003). More specifically, teachers need to undergo inquiry-type experiences similar to their own students According to Tamir (1989), short-term professional development experiences cannot provide teachers with the know-how to adequately operate in the more demanding student-centered learning environment in which they may face unforeseen and unplanned situations. Marx *et al.* (1998) observed that in the laboratory, teachers often face difficulties in helping students formulate insightful questions, design investigations, and draw conclusions.

To implement learning by inquiry in the science classroom and laboratory, it is essential that teachers have first-hand experience with all of the cognitive dimensions and practical stages where such skills are implemented. This includes asking relevant questions, handling and solving unforeseen problems, designing experimental conditions to resolve research questions, working in small collaborative groups, and conducting discussions. To sum up, teachers must switch from the "teaching-by-telling" instructional method to listening to students' ideas and questions.

7.4.1   The Inquiry Chemistry Laboratory Program in Israel

A series of 100 inquiry-type experiments were developed and incorporated into the regular chemistry curriculum used in grades 11 and 12 in the inquiry chemistry laboratory in Israel, most of them dealing with the key concepts taught in chemistry, *e.g.*, acids–bases, stoichiometry, oxidation–reduction, bonding, energy, chemical equilibrium, and reaction rates. The inquiry experiments range from totally 'open-ended' investigations to those in which the student is asked to conduct only a 'partial inquiry'. Some experiments include design and planning as well as interpretation of results and scientific conclusions, whereas others also include suggested hypotheses and relevant research questions (for more details, see Hofstein *et al.*, 2004). The teachers choose approximately 20 experiments by focusing on certain skills based on their teaching objectives and their students' abilities and interests.

The inquiry experiments are usually performed in small groups, 3–4 students per group. The number of groups can vary from four to eight, according to the total number of students in the class. During the inquiry stage of the experiment, each group of students usually defines a different inquiry question, conducts a different experiment, and may reach different conclusions. Each group summarizes its inquiry-type experiment in a document fondly termed a "Hot report", and hands it to the teacher immediately after the lesson for assessment. An example of the different phases is presented in Table 7.2.

The different groups are not closely involved with each other's experiments. If teachers want their students to share the knowledge each group has accumulated, and if they want to emphasize and strengthen certain issues concerning the experiment, they should conduct a summarizing discussion. Knowledge can refer to an inquiry skill and/or chemistry content knowledge. Because the inquiry approach is student-centered, most teachers start the summarizing discussion with students' presentations. For the students, presenting their work requires adequate preparation. The teacher should guide them in the way in which their work should be presented as well as in what aspects should be discussed.

7.4.2   The Chemistry Teacher's Practice in the Inquiry Chemistry Laboratory

Practice in science teaching can be defined in terms of the knowledge that teachers need in their teaching (Shulman, 1992; Magnusson *et al.*, 1999). According to Hofstein *et al.* (2004), accomplished teachers who are involved in this program should retain the following skills:

- Encourage students to interact professionally, including sharing knowledge with their peers, community members, or experts.
- Help students solve problems, ask high-level questions, and hypothesize regarding certain unsolved experimental problems.

Table 7.2 Phases in an inquiry-type experiment.

Phases in an example inquiry experiment	Abilities and skills
Phase 1: pre-inquiry	
• Insert the two solids, A and B, into the plastic bag and mix them by shaking	• Conducting an experiment using teacher's instructions
• Pour 10 ml of water into the small glass	
• Put the glass with the water inside the bag (be careful to avoid any contact between the water and the solids)	• Observing and recording observations
• Put a thermometer inside the bag to measure the temperature of the solids	
• Tie the bag carefully at its upper part (the thermometer is in the bag)	
• Turn over the glass and let the water completely wet the solids	
• Record all of your observations and answer the enclosed questionnaire	• Asking questions and hypothesizing
Phase 2: inquiry	
• Plan an experiment to investigate the question	• Planning an inquiry experiment
• Present a plan to conduct an experiment	
• Ask the teacher to provide you with equipment and materials to conduct the experiment	
• Conduct the experiment that you proposed	• Conducting the planned experiment and recording observations
• Observe and clearly note your observations	• Analyzing the results, asking further questions, and presenting the results in a scientific way
Phase 3: post-inquiry	
• Discuss with your group whether your hypothesis was accepted or must be rejected	

- Assess students continuously using a variety of alternative assessment methods.
- Customize the new activities according to their needs, and make decisions regarding the level of inquiry that is suitable for their students.
- Align the experiment with the concept taught or discussed in the chemistry classroom.
- Instruct students on how to observe chemistry phenomena.

To use the inquiry approach, teachers need to undergo an intensive process of professional development so that they will experience the same skills, knowledge, and thinking habits as their students (Windschitl, 2003). Moreover, in a CPD project, they should also undergo the entire inquiry process, so that they will be able to instruct and scaffold their students effectively. Teachers should gain adequate pedagogical content knowledge to serve as good guides for their students (Hofstein *et al.*, 2005). In addition, they should change their teaching strategies and adopt new ones, such as guiding instead of telling, supporting and encouraging students' interests, encouraging curiosity, accepting students' ideas, and responding to students' questions with new strategies rather than immediately giving them the answers.

Clearly, the laboratory classroom is very demanding. Thus, teachers need to undergo intensive and comprehensive professional enhancement. In addition, for many teachers, the inquiry approach in the chemistry laboratory is new and demands a radical change in their teaching and classroom behavior style. Inquiry-type experiences in the chemistry laboratory mean changing from a teacher-centered to a student-centered approach.

To understand the complications of teaching in the chemistry laboratory, we consider two approaches to the same practical experience: a traditional confirmatory laboratory using "recipe" style instructions and an inquiry-oriented approach. These two approaches differ in the following aspects:

- the roles of the teacher and the students
- skills that are used and developed during the practical activity
- time allotted and provision of equipment and materials
- open-endedness of the activity (different levels of inquiry, *e.g.*, guided by the teacher).

It should be recognized that these two approaches are at opposite ends of a continuum of practice in which teachers are able to draw on features from both when constructing their lessons, and as such, there are many available variations for laboratory work (Hofstein *et al.*, 2013). Thus, it is suggested that professional development of chemistry teachers to teach in the laboratory classroom learning environment should provide them with tools to design activities based on the above list of aspects. Teachers should be equipped with the ability to design experiments that will be aligned with the facilities they possess, students' abilities and interests, safety rules, the implementation of assessment methods aligned with pedagogical procedures, effective use of information and communication technology (ICT) in the context of the chemistry laboratory (for more about the use of ICT in the chemistry laboratory, see Chapter 8 of this book), and finally the teachers' own goals for conducting a certain practical activity. These skills, it is suggested, should be part of every pre- and in-service professional development initiative.

7.4.3   Organizing the Work in the Laboratory Classroom

As already noted, there are different ways of organizing laboratory work in a chemistry laboratory classroom. The most common organizational structures are the demonstration and the experiment, conducted cooperatively or individually by the students. The advantage of a demonstration is that the teacher can provide a step-by-step explanation of the experiment and the purpose of each individual activity, and he/she can focus the students' attention on the observations. On the other hand, in general, a demonstration makes the students passive recipients of information and thus only rarely are they are engaged in cognitive challenges.

Doing experiments in parallel has the advantage of turning the students themselves into active learners. But as discussed earlier in this chapter, such engagement only goes beyond the physical activity if the experiment allows for and challenges students' freedom of thought, *i.e.*, through inquiry. Just following a prescribed protocol (in confirmatory-type experiments) may lead to physical activity. But, moving to inquiry and open laboratory tasks also activates students' thought processes. Difficulties in parallel groups often appear when the amount of material is limited, such that the experiment can only be done in parallel a few times. It should be noted that in cases in which the experiment is very complex, inexperienced groups of learners might find it too demanding and thus difficult to perform on their own. Such experiments call for adequate and precise preparations by the teacher and/or the laboratory assistant. This might ensure that the students will obtain the correct solutions and experimental observations.

An important aspect of teachers' professional development is the ability to reflect on their own work, collect artifacts from their laboratory classroom, and construct evidence-based portfolios. *Portfolios* have been defined in different ways, depending on their purpose, which could include certification and selection, appraisal and promotion, or the teachers' CPD. The portfolio used for CPD purposes can include materials and samples of work that provide evidence for critical examination of teaching and learning practices (Klenowski, 2002). A critical aspect of portfolio development, initially recognized by Shulman (1992), is the importance of discussing teaching and learning with colleagues. Since then, other authors have noted the importance of sustained discussion and the use of teams to support the portfolio-development process (Davis and Honan, 1998). Grant and Huebner (1998) suggested that the portfolio should include a reflective commentary, the result of deliberation and conversations with colleagues, which allows others to examine the pedagogical decisions underlying the documented teaching.

7.5 Professional Development of Teachers to Teach in the Inquiry Chemistry Laboratory: An Evidence-based Approach

An application of the above practical and theoretical perspectives was implemented in Israel and demonstrated in a chemistry laboratory program entitled: *Learning in the chemistry laboratory by the inquiry approach* (Hofstein *et al.*, 2004) developed at the Department of Science Teaching of the

Weizmann Institute of Science. For this program, about 100 inquiry-type experiments were developed and implemented in 11th- and 12th-grade chemistry classes in Israel. A two-phase teaching process was used that consisted of a guided pre-inquiry phase followed by a more open-ended inquiry phase. Implementation of this program was very demanding for both the chemistry teachers and those who coordinated the development of the inquiry-type chemistry experiments, and the professional development of the teachers. An evidence-based professional development model was developed (for more about evidence-based professional development, see Hofstein, 2005; Harrison *et al.*, 2008). Figure 7.1 is a graphical presentation of the three phases that were part of the CPD.

7.5.1   Development of the CPD Model

The evidence-based CPD model was developed by the chemistry group in the Department of Science Teaching at the Weizmann Institute as part of a more comprehensive and collaborative CPD project that was conducted between King's College, London and the Weizmann Institute of Science (Hofstein, 2005: Taitelbaum *et al.*, 2008). This CPD model, which was tried out using a research design approach (Fortus *et al.*, 2004), was designed and implemented for a period of 3 years. The first year focused mainly on developing a teacher's guide and planning a summer induction course. The CPD model was implemented in the second and third years; the model was modified between the second and third year. Seven high-school chemistry teachers participated in this program each year. They were novices in teaching chemistry in the laboratory using the inquiry approach, but most of them had several years of experience in teaching chemistry. Here, we refer to the second and third year. The model consisted of three phases (Figure 7.1): (i) development of the CPD components, (ii) summer induction course, (iii) the workshop, which included preparation of an evidence-based portfolio and videotaped observations in the classroom laboratory at school aiming to providing the teachers with continuous support.

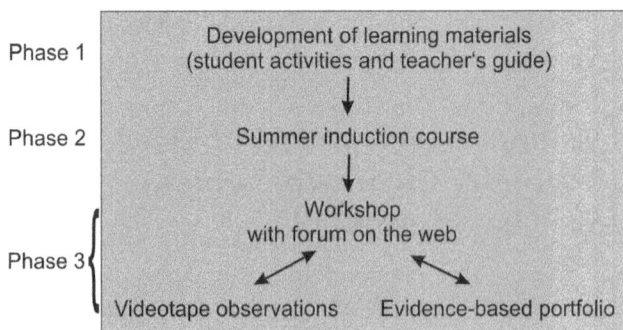

Figure 7.1   The three phases of the CPD model.

The teacher's guide was developed by a group of eight experienced teachers who had been teaching according to the inquiry approach for several years, in order to give some specific guidance to novice teachers concerning this approach, *e.g.*, guidelines for conducting experiments in the classroom laboratory (objectives, students' prerequisite knowledge, the position of the activity in the teaching sequence, the specific timeframe, specific safety notes, tips based on other teachers' experiences, examples of inquiry questions asked by students, and the scientific background needed for the experiment).

The main objective of the summer induction course was to provide novice teachers with an initial exposure to the basic methods (and skills) of using the inquiry approach in the laboratory. The summer induction course consisted of 8 hours a day for 5 days. During the induction course, the teachers learned and practiced inquiry skills by getting first-hand experience in all of the cognitive dimensions and the practical stages that accompany such learning, while assuming the role of students. The teachers worked in small collaborative groups, asked relevant questions, handled and solved unforeseen problems, and designed and conducted experiments to resolve research questions. They were exposed to the idea that the inquiry approach is a student-centered activity in which the teacher acts as a guide rather than just teaching facts. As already noted, the students' group work is a central feature of the inquiry approach. In the summer induction course, the various ways of grouping the students were discussed and practiced. The teachers conducted several experiments, each with a different set of peers in the group, determined by the course providers. It was recommended to change the composition of the groups for each activity according to various criteria, *e.g.*, heterogeneous grouping, homogeneous grouping, high achievers and low achievers, as well as according to gender.

The aim of the workshop was to select a group of teachers who were inexperienced in using the inquiry approach to share their experiences with their peers. In the workshop, they brought artifacts from their classes, reflected on their teaching, and summarized it by recording evidence. It should be noted that to transform an artifact into evidence, the teachers should be provided with opportunities to analyze the artifacts and also to reflect upon them. The workshop was conducted throughout the school year and consisted of seven monthly 3-hour meetings. Each meeting consisted of three parts. In the first part, the CPD providers for the workshop raised issues such as: what is reflection, an artifact, evidence, and why should a portfolio be constructed? In the second part, the teachers provided information about activities from their classroom laboratories, brought artifacts to support their presentations, and presented the procedure for transforming them into evidence. The artifacts could be students' "Hot reports", teachers' worksheets, or part of a videotape taken during certain activities. The presentation was followed by a discussion regarding the rationale for selecting the particular activity, the goals for conducting the particular experiment, and the teachers' behavior or teaching strategies used in the activity. The third part dealt with challenges and actual problems related to implementing the inquiry experiments in the classroom laboratory.

The aim of constructing an evidence-based portfolio was to enable the teachers to demonstrate their professional development during the school year, while reflecting and supporting it with evidence from their practice. Each teacher was asked to construct a portfolio that included three written pieces of evidence. Together with the teachers, we defined the types of items that could be included in a portfolio. The portfolio consisted of the following:

1. Introduction – presentation of the subject "teaching by the inquiry approach in the chemistry classroom laboratory" from the teacher's point of view.
2. Professional background.
3. Evidence from classroom laboratory activities – three pieces of evidence based on classroom laboratory teaching and specific artifacts taken from the classroom.
4. Reflective summary of the CPD.

To continue discussing issues that were raised during the workshop meetings, we set up an online closed internet forum that served only our group of teachers. During the school year, one of the authors observed and videotaped some classes several times. The videotape observations were intended to serve as a research tool. However, because videotaping of the activities was recommended for use as an artifact, the teachers used vignettes of the videos for their evidence and presentations.

Some examples of new pedagogical knowledge and content knowledge developed by the teachers were presented: the change in grouping the students, the change in managing the laboratories' lessons, *e.g.*, student-centered rather than teacher-centered, and the change in phrasing an inquiry question. The study was aimed at understanding some of the unique teaching strategies that teachers have to adopt while teaching the inquiry approach in their classes, and how these are developed and enriched throughout their various experiences. The study was based on the teachers' point of view, *e.g.*, the different opportunities for reflection, and was supported by looking into their practice, *e.g.*, the observations. Although we observed various teaching strategies in the classroom laboratory, we decided to focus on group work because the literature indicates its importance in achieving the teaching and learning goals. The results indicated that teachers had to develop their management of group work. A recommended way of doing this was by reflecting upon the preparations and the enactment of the inquiry activity.

It was suggested that during the CPD initiative, the teachers gained more self-confidence to criticize their own work and to understand their teaching strategies in leading and tutoring students who work in small collaborative groups, or to develop the investigative skills of students, such as discussing the types of questions posed, the nature of the hypothesis raised, the questions selected for further investigation, and the process of planning more experiments. Introducing evidence from the classroom laboratory was not a trivial task, but the fact that the teachers were encouraged to document their work, together with the process of investigating it during the classroom laboratory activity (reflecting and watching the videos), significantly contributed to their work. Among the benefits, the teachers mentioned the exchange of ideas, as well as getting relevant feedback and support. The model implemented in this study can be adopted effectively for other instructional techniques and pedagogical interventions used by science teachers in general and by chemistry teachers in particular. Supporting teachers continuously (as in this study or by teacher leaders) has the potential to enhance teachers' professional practice in an attempt to attain new and higher pedagogical standards.

7.6   Summary

- In this chapter, a genuine attempt was made to screen the science education literature with the aim of showing that the science (in our case chemistry) laboratory is a unique learning environment with respect to both teaching and learning.
- We show that as concerns the chemistry teacher, this pedagogy can be very demanding in terms of guiding the students and attaining goals of inquiry and other high-order learning skills. If we want the chemistry laboratory to be an effective learning platform, we need to effectively provide the teachers with a toolbox to act effectively in such practical activities.
- We show that one needs a CPD model that will be highly aligned with the goals, structure and students' skills. This CPD takes time and is quite challenging.
- In recent years, the development of learning skills has become as important as learning concepts and chemistry topics. The chemistry education community thus needs to develop effective CPD procedures that will eventually help attain these rather ambitious goals.
- Finally, in the past, the chemistry laboratory meant manipulating apparatus and chemicals. As we enter the 21st century, we need to research and develop a broader spectrum of goals for the chemistry laboratory in teaching.

References

Abrahams I. and Millar R., (2008), Does practical work really work? A study on the effectiveness of practical work as a teaching and learning method in school science, *Int. J. Sci. Educ.*, **30**, 1945–1969.

Barnett J. and Hodson D., (2001), Pedagogical content knowledge: Towards a fuller understanding of what good science teachers know, *Sci. Educ.*, **85**, 426–453.

Bates G. R., (1978), The role of the laboratory in secondary school science programs, in Rowe M. B. (ed.), *What Research Says to the Science Teacher*, Washington, DC: National Science Teachers Association (NSTA), Vol. 1.

Bennet J., (2006), Teaching and learning science: a guide to recent research and its application, London: Continuum.

Bybee R., (2000), Teaching science as inquiry, in Minstrel J. and Van Zee E. H. (ed.), *Inquiring into Inquiry Learning and Teaching*, Washington, DC: AAAS, pp. 20–46.

Chinn C. A. and Malhotra B. A., (2002), Epistemologically authentic inquiry in schools: a theoretical framework for evaluation inquiry tasks, *Sci. Educ.*, **86**, 175–218.

Davis C. L. and Honan E., (1998), Reflections on the use of teams to support the portfolio process, in Lyons N. (ed.), *With Portfolio in Hand: Validating the New Professionalism*, New York: Teachers College Press, pp. 90–102.

De Carlo C. L. and Rubba P. A., (1994), What happens during high school chemistry laboratory sessions? A descriptive case study of the behaviors exhibited by three teachers and their students, *J. Sci. Teach. Educ.*, **5**, 37–47.

Domin D. S., (1999), A content analysis of general chemistry laboratory manuals for evidence of higher-order cognitive tasks, *J. Chem. Educ.*, **76**, 109–111.

Driver R., Newton P. and Osborne J., (2000), Establishing the norms of scientific argumentation in classrooms, *Sci. Educ.*, **89**, 287–312.

Duschl R. A. and Osborne J., (2002), Supporting and promoting argumentation discourse in science education, *Stud. Sci. Educ.*, **38**, 39–72.

Fortus D., Dershimer R. C., Krajcik J., Marx R. W. and Mamlok-Naaman R., (2004), Design-based science (DBS) and student learning, *J. Res. Sci. Teach.*, **41**, 1081–1110.

Grant G. E. and Huebner T. A., (1998), The portfolio question: the power of self-directed inquiry, in Lyons N. (ed.), *With Portfolio in Hand: Validating the New Professionalism*, New York: Teachers College Press, pp. 156–171.

Gunstone R. F. and Champagne A. B., (1990), Promoting conceptual change in the laboratory, in Hegarty-Hazel E. (ed.), *The Student Laboratory and the Science Curriculum*, London: Routledge, pp. 159–182.

Harrison C., Eylon B., Hofstein A. and Simon S., (2008), Evidence-based professional development of science teachers in two countries, *Int. J. Sci. Educ.*, **30**, 577–591.

Hashweh M. Z., (1996), Effects of science teachers' epistemological beliefs in teaching, *J. Res. Sci. Teach.*, **33**, 47–64.

Hodson D., (1993), Re-thinking old ways: toward a more critical approach to practical work in school science, *Stud. Sci. Educ.*, **22**, 85–142.

Hofstein A., (2005), Evidence-based continuous professional development (CPD) programmes in six science domains, paper presented at the meeting of the National Association for Research in Science Teaching, San Francisco, CA, USA.

Hofstein A. and Kind P. M., (2012), Learning in and from science laboratories, in Fraser B., Tobin K. and McRobbie C. (ed.), *Second International Handbook of Science Education*, Dordrecht: Springer, pp. 189–209.

Hofstein A., Kipnis M. and Abrahams I., (2013), How to learn in and from the chemistry laboratory, in Eilks I. and Hofstein A. (ed.), *Teaching Chemistry – A Studybook*, Rotterdam: Sense Publishers, pp. 153–183.

Hofstein A. and Lunetta V. N., (1982), The role of the laboratory in science teaching: neglected aspects of research, *Rev. Educ. Res.*, **52**, 201–217.

Hofstein A. and Lunetta V. N., (2004), The laboratory in science education: foundations for the twenty-first century, *Sci. Educ.*, **88**, 28–54.

Hofstein A., Navon O., Kipnis M. and Mamlok-Naaman R., (2005), Developing students' ability to ask more and better questions resulting from inquiry-type chemistry laboratories, *J. Res. Sci. Teach.*, **42**, 791–806.

Hofstein A., Shore R. and Kipnis M., (2004), Providing high school chemistry students with opportunities to develop learning skills in an inquiry-type laboratory: a case study, *Int. J. Sci. Educ.*, **26**, 47–62.

Hurd P. D., (1983), Science education: the search for new vision, *Educ. Leadership*, **41**, 20–22.

Jimenez-Aleixandre M. P., Rodriguez A. B. and Duschl R. A., (2000), "Doing the lesson" or "doing science": argument in high school genetics, *Sci. Educ.*, **84**, 757–792.

Johnstone A. H. and Al-Shuaili A., (2001), Learning in the laboratory; some thoughts from the literature, *Univ. Chem. Educ.*, **5**(2), 42–91.

Johnstone A. H. and Wham A. J. B., (1982), The demands of practical work, *Educ. Chem.*, **19**(3), 71–73.

Katchevich D., Hofstein A. and Mamlok-Naaman R., (2013), Argumentation in the chemistry laboratory: inquiry and confirmatory experiments, *Res. Sci. Educ.*, **43**, 317–345.

Kind P. M., (2003), TIMSS puts England first on scientific enquiry, but does pride come before a fall? *Sch. Sci. Rev.*, **85**(311), 83–90.

Kind P., Wilson J., Hofstein A. and Kind V., (2011), Peer argumentation in school science laboratory – exploring effect of laboratory task feature, *Int. J. Sci. Educ.*, **33**, 2527–2558.

Kipnis M. and Hofstein A., (2008), The inquiry laboratory as a source for development of metacognitive skills, *Int. J. Sci. Math. Educ.*, **6**, 601–627.

Klenowski V., (2002), *Developing Portfolios for Learning and Assessment*, London: RoutledgeFalmer.

Kuhn D., (1988), Thinking as argument, *Harv. Educ. Rev.*, **62**, 155–178.

Kuhn D., Black J., Keselman A. and Kaplan D., (2000), The development of cognitive skills to support inquiry learning, *Cognit. Instr.*, **18**, 495–523.

Lazarowitz R. and Tamir P., (1994), Research on using laboratory instruction in science, in Gabel, D. L. (ed.), *Handbook of Research on Science Teaching*, New York: Macmillan, pp. 94–127.

Livingstone A., (1997), *Metacognition: An Overview Unpublished Manuscript*, Buffalo: State University of New York. Retrived from: http//www.gse. buffalo.edu/fas/sheuell/cep564/metacog.htm.

Lunetta V. N. (1998), The school science laboratory: Historical perspectives and centers for contemporary teaching, in Fraser B. J., Tobin Y. K. G. (ed.), *International Handbook of Science Education*, Dordrecht: Kluwer, pp. 249–262.

Lunetta V. N., Hofstein A. and Clough M. P. (2007), Learning and teaching in the school science laboratory: An analysis of research, theory and practice, in Abell S. K. and Lederman N. G. (ed.), *International Handbook of Science Education*. Mahwah, NJ: Lawrence Erlbaum, pp. 393–441.

Magnusson S., Krajcik J. and Borko H., (1999), Nature, sources, and development of pedagogical content knowledge for science teaching, in Gess-Newsome J. and Lederman N. G. (ed.), *Pedagogical Content Knowledge and Science Education*, Dordrecht: Kluwer Academic Publishers, pp. 95–132.

Marx R. W., Freeman J. G., Krajcik J. S. and Blumenfeld P. C. (1998), Professional development of science teachers, in Fraser B. J. and Tobin K. G. (ed.), *International Handbook of Science Education*, Dordrecht: Kluwer, pp. 249–262.

NRC (National Research Council), (1996), *National Science Education Standards*, Washington, DC: National Academy Press.

Rickey D. and Stacy A. M., (2000), The role of Metacognition in learning chemistry, *J. Res. Sci. Teach.*, 77, 915–920.

Sandoval W. A., (2005), Understanding students' practical epistemologies and their influence on learning through inquiry, *Sci. Educ.*, 89, 634–656.

Scott P., (1998), Teacher talk and meaning making in science classrooms a Vigotskian analysis and review, *Stud. Sci. Educ.*, 32, 45–80.

Sere G. M., (2002), Towards renewed research questions from outcomes of the European lab-work in science education, *Sci. Educ.*, 86, 624–644.

Shulman L., (1992), Portfolios in teacher education: a component of reflective teacher education, paper presented at the annual meeting of the American Educational Research Association, San Francisco, CA.

Solomon J., Duveen J. and Scott L., (1994), Pupils images of scientific epistemology, *Int. J. Sci. Educ.*, 16, 361–373.

Taitelbaum D., Mamlok-Naaman R., Carmeli M. and Hofstein A., (2008), Evidence for teachers' change while participating in a continuous professional development programme and implementing the inquiry approach in the chemistry laboratory, *Int. J. Sci. Educ.*, 30, 593–617.

Tamir P., (1989), Training teachers to teach effectively in the science laboratory, *Sci. Educ.*, 73, 59–69.

Tobin K. G., (1990), Research on science laboratory activities : In pursuit of better questions and answers to improve learning, *Sch. Sci. Math.*, 90, 403–418.

Windschitl M., (2003), Inquiry project in science teacher education: what can investigative experiences reveal about teacher thinking and eventual classroom practice? *Sci. Educ.*, 87, 112–143.

Woolnough B. E. and Alsop T., (1985), *Practical Work in Science*, Cambridge: University Press.

CHAPTER 8

Continuous Professional Development of Chemistry Teachers to Incorporate Information and Communication Technology

There is no doubt that the world of tomorrow will be much more permeated by the various types of media than today, among them all of the digital information and communication technologies (ICT). The term ICT describes the integration of telecommunication, computers, software and audiovisual systems to allow users to create, access, store, transmit, and manipulate any type of information (Stevenson, 1997). Most schools in developed countries already use computers, tablet PCs and other digital devices, and have access to the Internet. However, teachers' beliefs and conceptions about how to make the best use of modern ICT for chemistry teaching and how to cope with the corresponding challenges in chemistry education are continually changing. This chapter reflects the influence of modern ICT on chemistry teaching and the corresponding challenges for teacher education and continuing professional development (CPD).

8.1 Scientific Literacy, Media Literacy, and ICT

The PISA Framework (OECD, 2013), outlined by the Organisation for Economic Co-operation and Development (OECD), has identified scientific literacy as the overarching aim of all science education. Scientific literacy has been defined as "the ability to engage with science-related issues, and with

Advances in Chemistry Education Series No. 1
Professional Development of Chemistry Teachers: Theory and Practice
By Rachel Mamlok-Naaman, Ingo Eilks, George Bodner and Avi Hofstein
© Rachel Mamlok-Naaman, Ingo Eilks, George Bodner and Avi Hofstein 2022
Published by the Royal Society of Chemistry, www.rsc.org

the ideas of science, as a reflective citizen" (p. 7). A scientifically literate person is expected to be both able and willing to engage in reasoned discourse about science and technology. The PISA Framework suggests that this requires a certain set of skills:

- *Explain phenomena scientifically*: recognize, offer, and evaluate explanations for a range of natural and technological phenomena.
- *Evaluate and design scientific inquiry*: describe and appraise scientific studies and/or experiments and propose ways of addressing questions scientifically.
- *Interpret data and evidence scientifically*: analyze and evaluate data, claims, and arguments in a variety of representations and draw appropriate scientific conclusions.

Today's society is faced with natural and technological phenomena, scientific investigations, data, claims and arguments that come to us mainly *via* the media. Discussions about science and technology are no longer affected by face-to-face communication or one-way communication offered by traditional print media, TV, radio, or movies. Today, debates over scientific questions and socio-scientific issues take place interactively through bi- and multidirectional communication on the Internet, blogs, forums, and other types of social media. This means that investment is needed in teachers' understanding of contemporary media and communication technologies in general, and with respect to chemistry teaching in particular. Pedagogically reflected digital media literacy needs to be developed among chemistry teachers to allow the next generation to learn how to gain informed access to discourse about modern applications of chemistry and technology in the media.

Scientific literacy in the modern world is a multidimensional construct that includes media literacy and skills in the use of ICT (Rodrigues, 2010). Media literacy is one of the central cross-curricular goals in modern society, with many implications for reform in education. Today, this has to include literacy and skills in the use of digital media (Belova *et al.*, 2017). The special role of media and information literacy (MIL) in modern society in general, and with respect to digital media in particular, has also been extensively discussed by UNESCO, which suggests that MIL will be a key skill set in the 21st century:

MIL recognizes the primary role of information and media in our everyday lives. It lies at the core of freedom of expression and information – because it empowers citizens to understand the functions of media and other information providers, to critically evaluate their content, and to make informed decisions as users and producers of information and media content (http://www.unesco.org/new/en/communication-and-information/media-development/media-literacy/mil-as-composite-concept/).

The UNESCO MIL Framework, summarized in Table 8.1, assumes that teachers play a key role in making the young generation literate in the use of

Table 8.1 The UNESCO MIL curriculum framework for teachers.

Curriculum dimensions

Key curriculum areas	Knowledge of media and information for democratic discourse	Evaluation of media and information	Production and use of media and information
Policy and vision	Preparation of media- and information-literate teachers	Preparation of media- and information-literate students	Fostering of media- and information-literate societies
• Curriculum and assessment	Knowledge of media, libraries, archives and other information providers, their functions and the conditions needed to perform	Understanding of criteria for evaluating media texts and information sources	Skills to explore how information and media texts are produced, social and cultural context of information and media production; uses by citizens; and for what purposes
• Pedagogy	Integration of media and information in classroom discourse	Evaluation of content of media and other information providers for problem-solving	User-generated content and use for teaching and learning
• Media and information	Print-based media – newspapers and magazines; information providers – libraries, archives, museums, books, journals, *etc.*	Broadcast media – radio and television	New media – internet, social networks, delivery platform (computers, mobile phones, *etc.*)
• Organization and administration	Knowledge of classroom organization	Collaboration through MIL	Applying MIL to lifelong learning
• Teacher professional development	Knowledge of MIL for civic education, participation in the professional community and governance of their societies	Evaluation and management of media and information resources for professional learning	Leadership and model citizen; championing the promotion and use of MIL for teacher and student development

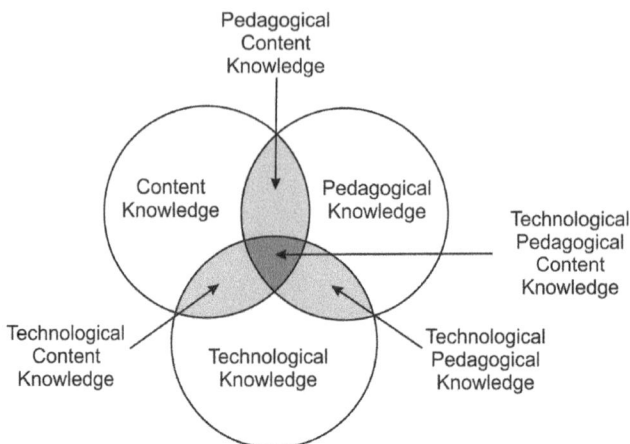

Figure 8.1 TPACK as a combination of content knowledge, pedagogical knowledge and technological knowledge (Mishra and Koehler, 2006).

media and information technologies, including both traditional media and more recently developed digital types of media. UNESCO therefore suggests a full MIL curriculum for teachers of all subjects to provide educators with the competency to integrate MIL into their classrooms (http://en.unesco.org/themes/media-and-information-literacy).

The UNESCO MIL approach is a general curriculum, not a domain-specific one. The question is whether this is enough, or whether we need more domain-specific approaches in addition to the development of general media and information skills (Krause *et al.*, 2017). This parallels the discussion about the importance of pedagogical content knowledge (PCK) as an addition to content knowledge (CK) and pedagogical knowledge (PK) (see Chapter 2 of this book). Such domain-specific knowledge about the use of ICT in education is called technological pedagogical content knowledge (TPACK). TPACK is a combination of CK, PK, and technological knowledge, as shown in Figure 8.1.

8.2 The "Traditional" Use of ICT in the Chemistry Classroom

Computers for teaching chemistry in secondary schools emerged in the 1980s (Lunetta, 1998). Compared to today, the capacity of those computers and the available software were very limited. The computers were neither networked, nor connected to any other information source outside the classroom. They were not able to store large amounts of data. In the beginning, they could not even handle the subscripts and superscripts needed to convey chemical information.

Nevertheless, computers provided new opportunities for teaching and learning chemistry. Early applications focused on digital data recording and evaluation. Interfaces to chemical sensors allowed monitoring chemical changes, *e.g.*, by pH and conductivity sensors, or photometers. Data were digitally collected and exported into programs, such as spreadsheets, to evaluate and display them. Today, new devices allow direct data collection by software that is designed to be compatible with sensors, such as the probes offered by Vernier, Bluetooth or Wi-Fi-based sensors. These sensors can easily be connected to the students' smartphones and tablet PCs. Data are recorded by corresponding apps that often allow direct evaluation of the data and export of the corresponding tables and diagrams. However, these de-velopments mandate CPD programs for teachers because the available technologies are rapidly and continuously changing.

Another early application aimed at better visualization in the form of three-dimensional representations of molecules and lattice structures, and visual animations of chemical processes (Barnea and Dori, 1999). Recent developments have made access to the corresponding tools and information even easier. Instant online access to large databases enables viewing a vast and continuously growing number of molecular structures. Thanks to easy and intuitive software, learners can draw chemical structures, optimize them based on quantum-mechanical models, and display them in different representational modes. New software also facilitates integrating new approaches to understanding chemical structures, such as symmetry con-siderations (Tuvi-Arad and Blonder, 2010).

It seems that these two approaches to using digital media in class, namely data collection and visualization, are the most frequently used applications of ICT in chemistry education. However, both technologies were, and in many ways still are, used more often by advanced students, *e.g.*, in advanced senior high-school courses or even undergraduate chemistry education.

A significant change in the use of ICT in chemistry education was initiated by the growth of the Internet in the 1990s (Dori *et al.*, 2013). The Internet provided a new way to access information on chemicals, chemical processes, applications and corresponding discussions, or on learning environments about them (*e.g.*, Tuvi and Nachmias, 2003). This became a challenge for teachers because with the traditional media, digital measurements or demonstrations, they had almost complete control over the students' access to chemistry-related information. Media used in class were mainly restricted to the textbook, worksheets, photocopies and information provided by the teacher. With the Internet, students now have access to different represen-tations of chemical content at any time, in and outside of class. This in-formation regularly includes alternative or even misleading interpretations or visualizations of chemical content (Eilks *et al.*, 2010). It also offers access to all kinds of pro and con arguments on each technological development related to chemistry – whether or not they are scientifically and technolo-gically justified.

8.3   Developing the School Science Laboratory by the Use of ICT

When computers appeared in schools in the 1980s, they were recognized as potentially useful tools to improve school science (Lunetta, 1998). Accumulated evidence showed that using appropriate technologies in the school science laboratory offers new ways to enhance the learning of scientific ideas. Inquiry-empowering technologies were developed and adapted to assist students in gathering, organizing, visualizing, interpreting, and reporting data (Hofstein and Lunetta, 2004). Some teachers and students also used new technology tools to gather data from multiple trials and over long time intervals (Friedler *et al.*, 1990; Lunetta, 1998; Krajcik *et al.*, 2000; Dori *et al.*, 2004). When teachers and students properly use inquiry-empowering technologies to gather and analyze data, students have more time to observe, reflect, and construct the conceptual knowledge underlying their laboratory experiences. Using appropriate ICT tools can enable students to conduct, interpret, and report more complete, accurate, and interesting investigations. Such tools can also provide media that support communication, student–student collaboration, the development of a community of inquirers in the laboratory classroom and beyond, and the development of argumentation skills (Zembal-Saul *et al.*, 2002). However, there is always the risk that digital measurements and data evaluation will become "black-box" technologies, resulting in a superficial understanding of the physical bases of the measurements.

Two early studies illustrated the potential effectiveness of particular technologies in high-school chemistry laboratories. In the USA, Nakhleh and Krajcik (1994) investigated how students' use of chemical indicators, pH meters, and microcomputer-based laboratories affects their understanding of acid–base reactions. Students who used computer tools in the laboratory emerged with a better ability to draw relevant concept maps, describe the acid–base construct, and argue about the probable causes of their graphs turning out as they did. In Israel, Dori *et al.* (2004) developed a high-school chemistry lesson plan in which the students pursued chemistry investigations using integrated desktop computer probes. In a pre-post design study, they found that students' experiences with the technology tools improved their ability to pose questions, use graphing skills, and pursue scientific inquiry more generally.

There is some evidence for the promise of integrating ICT tools in the science laboratory. The level at which ICT is used in the laboratory, however, varies a great deal from school to school. In the future, it is expected that ICT will cause more integration between practical work and computer-based simulations. Emerging technologies such as wireless sensors, cloud solutions for data exchange, and network tools for joint data evaluation will offer opportunities to develop chemistry education and create challenges for the professional development of chemistry teachers.

8.4 Current Challenges in the Use of ICT in Chemistry Education

The continuous expansion of the Internet and the emergence of wireless communication technologies, especially smartphones and tablet PCs, at the turn of the 21st century changed the world. In many industrial countries, more than 90% of students between 10 and 19 years of age own smartphones with wireless connection to the Internet. This development calls for a rethinking of the pedagogy of media use in chemistry education.

Various developments are expected to influence the practice of chemistry teaching in the future (Dori *et al.*, 2013). Current developments suggest that we will be able to integrate more and more of the real world using digital elements. Such virtual worlds might offer places to practice chemistry virtually, *e.g.*, training in a particular laboratory procedure in Second Life before performing it in a real-life laboratory (Winkelmann *et al.*, 2014). Augmented reality allows the combination of real science objects with additional digital content (Cai *et al.*, 2014). Multimedia learning books allow the integration of traditional text- and picture-based information with digital objects such as videos, visualizations and animations (Wilbraham *et al.*, 2012). Teachers will more often be asked to learn and develop better ways of using these technologies as part of their TPACK, in terms of both technology and pedagogy.

However, these new developments are not only expected to provide a richer classroom environment that consumes digital content in ICT-supported lessons. Developments in ICT will also provide richer opportunities for the teacher to create their own digital learning environments based on software such as PREZI (Krause and Eilks, 2014). Programs such as *iBook Author* even allow teachers to create their own non-linear and digitally supported Multi-Touch Learning eBooks for their students that can be individually adaptable and differentiated for use in heterogeneous learning groups (Huwer and Eilks, 2017). Such books can even be developed jointly with advanced students (Hill *et al.*, 2016).

Finally, a significant challenge for teachers' professional development as a result of the development of new digital media are the changes occurring in the ways of communication. A great deal of communication among young people now occurs through social media. Facebook, WhatsApp, YouTube, Internet forums, *etc.*, provide new ways of exchanging information and providing media coverage that could not exist using traditional media. Not much research has yet been done on how to use the newest types of digital media, *i.e.*, social media, to enrich pedagogy in science classes. A recent study on the use of Internet forums in junior high-school chemistry education in Germany showed some promising indications of social media being an effective tool to promote cooperative learning (Dittmar and Eilks, 2016). However, other research has shown that this will require intensive professional development for chemistry teachers to acquaint them with the necessary skills and self-efficacy (Blonder *et al.*, 2013).

8.5   Implications for Chemistry Teacher Education and Professional Development

Today's ever-changing media arena calls for continual renewal of the pedagogy of education in general and chemistry teaching in particular (Dori *et al.*, 2013). A recent study from Germany showed that the attitudes and self-efficacy of the young generation of prospective chemistry teachers in the use of ICT are quite promising (Krause *et al.*, 2017). However, that study also showed that investment is needed in developing teachers' self-efficacy in the use of ICT in chemistry education, especially among prospective female teachers. The study showed that domain-specific teacher education courses are more suitable for improving attitudes and self-efficacy in the use of ICT in chemistry classes than courses on the general use of ICT in teaching. This means that domain-specific practical courses on using ICT in chemistry teaching would be useful additions to pre-service chemistry teacher education programs. The courses need to show how to specifically use ICT for typical chemistry classroom activities. Studies on corresponding attitudes and self-efficacy among more experienced teachers are not available. However, taking into account that the previous generations of teachers did not grow up with digital media or the Internet, there is no sound reason to assume that the same courses are not needed for experienced teachers as well.

CPD in the use of ICT will be necessary, because chemistry teachers' knowledge of ICT will be outdated within a few years. Professional development needs to focus on general aspects of ICT, such as the availability of general devices – computer hardware and networks, for example – but also of specific devices such as new wireless chemical sensors or software for chemical structures, visualization, and animation. Professional development also needs to focus on the related pedagogical innovations for use of the corresponding devices and software in class – the TPACK.

According to the definition of media literacy by UNESCO and the MIL curriculum discussed at the beginning of this chapter, such courses should incorporate skills for using already existing media and for generating own media content. Constructivist theory suggests that this should be taught by a learner-centered approach. Professional development needs to incorporate activities that provide the teachers with experiences in the use of new devices and software to allow their use in their own classrooms. Teachers should also get a chance to create their own media, such as PREZI environments (Krause and Eilks, 2014), StopMotion videos (Krause and Eilks, 2017), or YouTube videos (Blonder *et al.*, 2013). With the emergence of new software and tools, this is getting progressively easier from a technical perspective; however, the pedagogical challenges for using the technology in a sound and helpful educational way still remain.

Recently, Krause and Eilks (2015) described a project focusing on the development of a course for pre-service chemistry teacher education in Germany on the use of ICT. The course consisted of six or seven 4-hour sessions. Using a cyclical, action research-based design, the course was developed from an initial idea to incorporate new ICT elements and corresponding educational strategies for chemistry education based on student feedback. Even after the project was completed, the course was subject to continuing changes and is still being updated, with new elements incorporated annually. Currently, the course comprises the following elements, which are introduced theoretically and experienced practically:

- Creating teaching materials with digitally generated molecular structures and experimental set-ups
- Creating chemistry learning environments with PREZI
- Designing WebQuests on chemistry-related topics
- Creating animated visualizations with StopMotion technology
- Using educational and chemistry-related apps for tablet PCs and smartphones
- Critically evaluating digital resources for chemistry education from the Internet
- Understanding and training for the function and handling of interactive whiteboards
- Using wireless and USB chemical sensors for digital data collection and evaluation
- Understanding legal issues in using materials from the Internet in chemistry teaching.

All elements of the course are also used in individual half-day courses that focus on certain aspects of the use of ICT in the chemistry classroom presented to in-service chemistry teachers. The courses are offered as teacher workshops or are run upon request for internal workshops organized by individual schools. One example is on using StopMotion apps to create small video sequences. Teachers are introduced to corresponding apps and develop their own videos as they would with their students. A recently described example involved learning about the nomenclature of organic compounds by creating StopMotion videos (Krause and Eilks, 2017). The teachers, like their students in class, think about naming organic molecules step by step according to IUPAC rules. StopMotion apps enable them to connect corresponding pictures into a video sequence. The sequences demonstrate how the IUPAC rules for the nomenclature of organic substances have to be applied. This teaching activity won first place in 2016 in the European *Contest on Digital Learning Objects in Chemistry* (Figure 8.2).

However, more promising programs to develop teachers' skills for using ICT in their classrooms are long term. For example, Eilks (2013) described a 5-year action research project to innovate the teaching and learning of the particulate nature of matter in lower secondary chemistry classes with the

Figure 8.2   Creation of a StopMotion video with an iPad attached to a flexible tripod as an activity in ICT-focused chemistry teacher CPD.

Figure 8.3   Students learning bionics based on a learning environment that was developed within a long-term professional development program as part of the PROFILES project.

integration of ICT (see Chapter 5 of this book). A group of chemistry teachers and a chemistry educator cooperatively developed digital learning environments and corresponding strategies on how to integrate them into junior high-school chemistry teaching. Accompanying research showed a substantial change in teachers' knowledge, skills, and attitudes (Eilks and Markic, 2011) that had a sustained effect on both the curriculum in the participating schools and the dissemination of the jointly developed practices (Eilks, 2014). Similar experiences were gained in the development of teaching and learning materials based on digital learning environments for bionics (Figure 8.3) as part of the PROFILES project. In 2014, this CPD project (Krause *et al.*, 2015) and the related lesson plans (Krause *et al.*, 2014a,b) were awarded the *STEM of Tomorrow School Award* in Germany.

8.6 Use of ICT for Professional Development of Chemistry Teachers

The professional development of chemistry teachers can use ICT not only as the focus, but also as the medium for the delivery of professional development. New technologies offer opportunities for teacher professional development beyond the limits of the corresponding technologies in class. Teachers can use ICT themselves to update their knowledge about current issues in chemistry, such as nanotechnology or Green Chemistry. The Internet offers a broad base of information that also includes teaching and learning materials, from worksheets to animations and videos. However, teachers need to be very critical. Teaching material offered on the Internet for chemistry education is not always accurate in terms of scientific correctness or helpful in the sense of didactical quality; some materials even contain well-known student misconceptions and learning difficulties (Eilks *et al.*, 2010).

ICT offers chances for teachers to work together to jointly develop their professional knowledge in teachers' learning communities (Vazquez-Abad *et al.*, 2004; see Chapter 4 of this book). ICT provides a useful format that allows learning communities to cooperate across districts or even countries. ICT also provides tools that facilitate exchange and communication or provide a basis for reflecting on teaching practices based on e-portfolios (Avraamidou and Zembal-Saul, 2003). Recently, Laudonia and Eilks (2018) described a case in which a vocational school chemistry teacher in a small city in southern Switzerland showed interest in cooperative chemistry curriculum development as part of his professional development. The curriculum development was based on participatory action research (PAR) as described for chemistry education by Eilks and Ralle (2002). The focus was on teaching chemical bonding in a student-active mode with a digital learning environment and self-assessments as suggested by Krause *et al.* (2013). Unfortunately, there was only one vocational school of this type in the local environment and no higher education institution to support the PAR process. The teacher contacted an already existing network of teachers who were experienced in PAR in the west of Germany supported by chemistry educators from the University of Bremen (Eilks, 2014). Over three consecutive cycles of curriculum development, the teacher was remotely embedded in an existing PAR network. Asynchronous and synchronous communication were available by a combination of email, Dropbox and Skype technologies. The experienced PAR network was able to support the teacher in his curriculum development and corresponding research. The teacher learned how to structure digital learning environments with PREZI (Krause and Eilks, 2014) and to create self-assessments for the students (Krause *et al.*, 2013). The teacher also developed skills in reflecting on and assessing his own teaching.

In the end, the teacher valued the remote embedding by ICT into a teacher network. However, this did not represent a one-directional flow of support and learning. The action research network benefitted from the process of reflecting on their own practices of teaching chemical bonding and gained access to the digital teaching and learning materials.

8.7   Implications

ICT provides both opportunities and challenges for chemistry education and professional development resulting from both rapid changes in ICT and the different experience levels of the people involved. More experienced teachers received their education before our contemporary ICT technologies became broadly available in schools. Other teachers grew up with computers but were educated as teachers before the revolution in ICT caused by expansion of the Internet and the emergence of handheld devices and social media. Younger teachers might have more common experiences and access to contemporary ICT, but even they cannot all be considered digital natives. However, this is the case for the students of today and the coming generations, at least in developed countries.

Keeping knowledge of the use of ICT in education up to date might be one of the biggest challenges for teachers' CPD in a world of rapidly changing media. Corresponding knowledge and skills include technological aspects, such as the technical operation and handling of new devices and software, and the specific use of ICT in the chemistry classroom. Continuing efforts are needed from the teachers to provide students with the best chance for effective learning by ICT.

Domain-specific approaches for teacher professional development are needed because the use of ICT in chemistry education is different from that in other subjects, as described in Chapter 2 of this book. The combination of knowledge, skills, understanding and competency within this domain is captured by the concept of TPACK. The unique nature of chemistry with its different representations, namely the macroscopic, submicroscopic, and formal representational levels, makes it a unique approach to understanding the world around us (Johnstone, 1991). Although ICT is a challenge, it also provides an opportunity. Chemistry education can specifically benefit from using ICT to visualize non-visible structures and processes at the atomic and molecular levels. Digital collection and evaluation of data generated in the laboratory can show students the power of modern chemistry.

New technologies provide ever-changing possibilities. Chemistry teaching and learning can be enriched by the more networked presentations and the possibility of exchanging any kind of information *via* the Internet and social media. To make the best use of them, teachers need to understand their students' media habits. However, they also need to know how information is handled in digital media to educate their students to become critical consumers of information about chemistry and technology on the Internet. This makes media education in chemistry part of relevant education for the future self-determined lives of students in a media-dominated society (Belova *et al.*, 2015).

8.8 Summary

- Modern ICT offers unique opportunities to support chemistry teaching and learning. ICT can help better integrate chemistry learning by incorporating its different representations, namely the macroscopic, submicroscopic, symbolic and contextual representational levels.
- Continuous updating of teachers' skills in the use of ICT for the teaching and learning of chemistry is among the biggest challenges for chemistry teacher professional development because of the continual advances in the corresponding technologies. Teachers need to continuously invest in their skills for operating current technological developments in ICT and in developing their knowledge about how to use ICT to support chemistry learning.
- Teachers need to keep their general knowledge up to date on both teaching and learning with ICT in general and in teaching chemistry in particular. Teachers need chemistry-specific TPACK to make the best use of current ICT in supporting the learning of chemistry among their students.

References

Avraamidou L. and Zembal-Saul C., (2003), Exploring the influence of web-based portfolio development on learning to teach elementary science, *J. Technol. Teach. Educ.*, **11**, 415–442.

Barnea N. and Dori Y. J., (1999), High-school chemistry students' performance and gender differences in a computerized molecular modeling learning environment, *J. Sci. Educ. Technol.*, **8**, 257–271.

Belova N., Dittmar J., Hansson L., Hofstein A., Nielsen J. A., Sjöström J. and Eilks I., (2017), Cross-curricular goals and the raise of the relevance of science education, in Hahl K., Juuti K., Lampiselkä J., Lavonen J. and Uitto A. (ed.), *Cognitive and Affective Aspects in Science Education Research*, Dordrecht: Springer, pp. 297–307.

Belova N., Stuckey M., Marks R. and Eilks I., (2015), Understanding the use of chemistry related information in the public, in Eilks I. and Hofstein A. (ed.), *Relevant Chemistry Education – From Theory to Practice*, Rotterdam: Sense, pp. 185–204.

Blonder R., Jonatan M., Bar-Dov Z., Benny N., Rapa S. and Sakhnini S., (2013), Can You Tube it? Providing chemistry teachers with technological tools and enhancing their self-efficacy beliefs, *Chem. Educ. Res. Pract.*, **14**, 269–285.

Cai S., Wang X. and Chiang F.-K., (2014), A case study of Augmented Reality simulation system application in a chemistry course, *Comp. Hum. Behav.*, **37**, 31–40.

Dittmar J. and Eilks I., (2016), Practical work, cooperative learning and Internet forums – an example on teaching about the chemistry of water, in Eilks I., Markic S. and Ralle B. (ed.), *Science Education Research and Practical Work*, Aachen: Shaker, pp. 239–244.

Dori Y. J., Rodrigues S. and Schanze S., (2013), How to promote chemistry learning through the use of ICT, in Eilks I. and Hofstein A. (ed.), *Teaching Chemistry*, Rotterdam: Sense, pp. 213–240.

Dori Y. J., Sasson I., Kaberman Z. and Herscovitz O., (2004), Integrating case-based computerized laboratories into high school chemistry, *Chem. Educ.*, **9**, 1–5.

Eilks I., (2013), Teachers' ways through the particulate nature of matter in lower secondary chemistry teaching: a continued change of different models vs. a coherent conceptual structure? in Tsaparlis G. and Sevian H. (ed.), *Concepts of Matter in Science Education*, Dordrecht: Springer, pp. 213–230.

Eilks I., (2014), Action Research in science education: from a general justification to a specific model in practice, in Stern T., Rauch F., Schuster A. and Townsend A. (ed.), *Action Research, Innovation and Change*, London: Routledge, pp. 156–176.

Eilks I. and Markic S., (2011), Effects of a long-term Participatory Action Research project on science teachers' professional development, *EurAsia J. Math. Sci. Technol. Educ.*, **7**(3), 149–160.

Eilks I. and Ralle B., (2002), Participatory Action Research in chemical education, in Ralle B. and Eilks I. (ed.), *Research in Chemical Education – What Does This Mean?* Aachen: Shaker, pp. 87–98.

Eilks I., Witteck T. and Pietzner V., (2010), Using multimedia learning aids from the Internet for teaching chemistry – not as easy as it seems? in Rodrigues S. (ed.), *Multiple Literacy and Science Education: ICTS in Formal and Informal Learning Environments*, Hershey: IGI Global, pp. 49–69.

Friedler Y., Nachmias R. and Linn M. C., (1990), Learning scientific reasoning skills in microcomputer-based laboratories, *J. Res. Sci. Teach.*, **27**, 173–191.

Hill A. C., Nickels L. M. and Sims P. A., (2016), Student-led development of an interactive and free biochemical methods eBook, *J. Chem. Educ.*, **93**, 1034–1038.

Hofstein A. and Lunetta V. N., (2004), The laboratory in science education: foundation for the 21st century, *Sci. Educ.*, **88**, 28–54.

Huwer J. and Eilks I., (2017), Multitouch Learning Books für schulische und außerschulische Bildung [Multitouch Learning Books for school and out-of-school education], in Meßinger-Koppelt J., Schanze S. and Groß J. (ed.), *Lernprozesse mit digitalen Werkzeugen unterstützen*, Hamburg: JHS Publishing, pp. 81–94.

Johnstone A. H., (1991), Why is science difficult to learn? Things are seldom what they seem, *J. Comp. Assisted Learn.*, **7**, 75–83.

Krajcik J., Blumenfeld B., Marx R. and Soloway E., (2000), Instructional, curricular, and technological supports for inquiry in science classrooms, in Minstell J. and Van Zee E. (ed.), *Inquiry into Inquiry: Science Learning and Teaching*, Washington: AAAS, pp. 283–315.

Krause M. and Eilks I., (2014), Innovating chemistry learning with PREZI, *Chem. Act.*, **104**, 19–25.

Krause M. and Eilks I., (2015), Lernen über digitale Medien in der Chemielehrerausbildung – Ein Projekt Partizipativer Aktionsforschung [Learning about digital media in chemistry teacher education – a project of participatory action research], *Chem. Kon.*, **22**, 173–178.

Krause M. and Eilks I., (2017), Learning about the nomenclature of organic substances by the creation of stop-motion videos, *Chem. Act.*, **109**, 36–38.

Krause M., Kienast S., Witteck T. and Eilks I., (2013), On the development of a computer-based learning and assessment environment for the transition from lower to upper secondary chemistry education, *Chem. Educ. Res. Pract.*, **14**, 345–353.

Krause M., Ostersehlt D. and Eilks I., (2015), Collaborative curriculum development of a teaching and learning module on bionics based on innovative ICT technology, in Yates N. L. (ed.), *New Developments in Science Education Research*, Hauppauge: Nova, pp. 51–64.

Krause M., Ostersehlt D., Mehrwald T., Runden H.-J., Mroske S. and Eilks I., (2014a), Using PREZI-Technology to promote guided inquiry learning on the topic 'bionics', in Bolte C., Holbrook J., Mamlok-Naaman R. and Rauch F. (ed.), *Science Teachers' Continuous Professional Development in Europe. Cases from the PROFILES Project*, Berlin: FU Berlin, pp. 149–158.

Krause M., Ostersehlt D., Meile J., Griemsmann L., Eilks I., Mehrwald T., Runden H.-J., Heering U. and Mroske S., (2014b), Using PREZI-technology to promote inquiry learning on bionics: two further modules, in Bolte C. and Rauch F. (ed.), *Enhancing Inquiry-based Science Education and Teachers' Continuous Professional Development in Europe: Insights and Reflections on the PROFILES Project and Other Projects Funded by the European Commission*, Berlin: FU Berlin, pp. 220–223.

Krause M., Pietzner V., Dori Y. J. and Eilks I., (2017), Differences in attitudes and self-efficacy of prospective chemistry teachers concerning the use of ICT in education, *EurAsia J. Math. Sci. Technol. Educ.*, **13**(8), 4405–4417.

Laudonia I. and Eilks I., (2018), Teacher-centred action research in a remote participatory environment – a reflection on a case of chemistry curriculum innovation in a Swiss vocational school, in Calder J. and Foletta J. (ed.), *(Participatory) Action Research (PAR): Principles, Approaches and Applications*, Hauppauge: Nova, pp. 215–231.

Lunetta V. N., (1998), The school science laboratory: historical perspectives and centers for contemporary teaching, in Fraser B. J. and Tobin K. G. (ed.), *International Handbook of Science Education*, Dordrecht: Kluwer, pp. 249–262.

Mishra P. and Koehler M. J., (2006), Technological pedagogical content knowledge: a framework for teacher knowledge, *Teach. Coll. Rec.*, **108**, 1017–1054.

Nakhleh M. B. and Krajcik J. S., (1994), The influence of level of information as presented by different technologies on students' understanding of acid, base, and pH concepts, *J. Res. Sci. Teach.*, **31**, 1077–1096.

OECD (2013). PISA 2015 draft science framework, retrieved www.oecd-ilibrary.org/education/how-does-pisa-assess-science-literacy_5jln4nfnqt7l-en.

Rodrigues S., (2010), *Multiple Literacy and Science Education: ICTs in Formal and Informal Learning Environments*, Hershey: IGI Global.

Stevenson D., (1997), *The Independent ICT in Schools Commission Information and Communications Technology in UK Schools, an Independent Inquiry*, London: The Independent ICT in Schools Commission.

Tuvi I. and Nachmias R., (2003), A study of web-based learning environments focusing on atomic structure, *J. Comp. Math. Sci. Teach.*, **22**, 225–240.

Tuvi-Arad I. and Blonder R., (2010), Continuous symmetry and chemistry teachers: learning advanced chemistry content through novel visualization tools, *Chem. Educ. Res. Pract.*, **11**, 48–58.

Vazquez-Abad J., Brousseau N., Guillermina W. C., Vezina M., Martinez A. D. and de Verjovsky J. P., (2004), Fostering distributed science learning through collaborative technologies, *J. Sci. Educ. Technol.*, **13**, 227–232.

Wilbraham A. C., Staley D. D., Matta M. S., Waterman E. L., (2012), *Pearson Chemistry*, London: Pearson.

Winkelmann K., Scott M. and Wong D., (2014), A study of high school students' performance of a chemistry experiment within the virtual world of Second Life, *J. Chem. Educ.*, **91**, 1432–1438.

Zembal-Saul C., Munford D., Crawford B., Friedrichsen P. and Land S., (2002), Scaffolding preservice science teachers' evidence-based arguments during an investigation of natural selection, *Res. Sci. Educ.*, **32**, 437–463.

CHAPTER 9

How to Educate Chemistry Teachers to Become Leaders

Conventional methods of conducting pre-service and in-service education and professional development are either inadequate or inappropriate for attaining the demanding goals of the new curriculum approaches. In-service workshops are too short and too infrequent to foster fundamental changes in teachers' classroom practice. To meet the challenges of reform in science education, we need to help schools and other educational institutions that are involved in this reform meet the challenges of standards-based approaches to teaching and learning. One way to progress toward these goals is to treat teachers as equal partners in the decision making regarding research and development. Where this is done, teachers will have to play a greater role in providing key leadership at all levels of the educational system.

9.1   Development of Leading Chemistry Teachers

The introduction of standards in science and mathematics education is often traced back to the AAAS (1993) *Benchmarks for Scientific Literacy* report. The science education standards as described by the National Research Council (1996) are suggestions for "best practices" that reflect the current vision of the content, classroom environment, teaching methods, and support necessary to provide a high-quality education in the sciences for all students. More recently, the Next Generation Science Standards (NGSS Lead States, 2013) have been developed to describe "best practices" for teaching K-12 science content in the USA. The NGSS approach calls for improving science education through learning across three dimensions.

Advances in Chemistry Education Series No. 1
Professional Development of Chemistry Teachers: Theory and Practice
By Rachel Mamlok-Naaman, Ingo Eilks, George Bodner and Avi Hofstein
© Rachel Mamlok-Naaman, Ingo Eilks, George Bodner and Avi Hofstein 2022
Published by the Royal Society of Chemistry, www.rsc.org

One of these dimensions is described as "cross-cutting concepts", such as *Cause and Effect: Mechanisms and Explanations*, which apply to the physical sciences, life sciences, earth and space sciences, and engineering. Another dimension is "science and engineering practices" that describe what scientists and engineers do to design and build systems. The third dimension is "disciplinary core ideas", which are key ideas, such as *Matter and its Interactions* and *Energy*, that build upon each other as students progress through the different grade levels, and are particularly relevant to the physical sciences. The content of the NGSS approach is research-based and it was developed by a collaboration between states in the USA for use by those states. Whereas the NGSS approach deals with the teaching and learning of science at the K-12 level, in general, the American Chemical Society (ACS) is involved in its application to the teaching of chemistry (Bodner, 2011). Clearly, these goals are very demanding and ambitious, requiring intensive and extensive cooperation between chemistry curriculum developers, chemistry teachers, and chemistry education researchers. The teachers who will eventually be involved in curricular initiatives are characterized as leading teachers.

Leadership among teachers has been defined as the ability of a person to bring about changes among teachers and in teaching (Fullan, 1991). In the context of science education, Pratt (2001) suggested that there are four basic skills relevant to effective leaders in science education: (i) technical, (ii) conceptual, (iii) interpersonal, and (iv) self-learning. Where teachers have been involved in decision making, curriculum development and implementation, and policy making, research has shown an increase in the retention of teachers and a net benefit to the communities in which these schools are located. In Chapter 3 of this book, we described the importance of involving teachers in the development and implementation of chemistry curricula for both continuous professional development and the development of leading teachers who will be eventually involved in the process of selecting the content and its related pedagogies.

Bybee (1993) suggested that developing leaders as well as curriculum developers among teachers is vital in an era of reforms in both the content of science teaching and the way in which science is taught. Research on professional development has shown that highly qualified leadership is required to foster changes in teaching and learning in schools (Fullan, 1991; Lawrenz, 2001). The studies described in this chapter suggest that the development of leadership is a demanding and complex process requiring a change in some aspects of intellectual activity. More specifically, according to Friel and Bright (1997), it requires explicit attention, clear expectations, and resources of both time and expertise. Our main goal in this chapter is to present and discuss the educational effectiveness of a model for the development of chemistry-teacher leaders and to assess the teachers' change process. It should be noted that this exploration is more descriptive than analytical in nature.

9.2   Models for the Development of Leadership among Chemistry Teachers: From Theory to Practice

This section describes an innovative program developed in Israel, at The National Center for Chemistry Teachers of The Weizmann Institute of Science, whose main goal was to improve the pedagogy of chemistry education in the Israeli educational system. It focuses on a model aimed at the professional development of chemistry-teacher leaders.

Israel has a centralized education system in which syllabi and curricula are regulated by the Ministry of Education. Since the 1960s, this ministry has provided for the long-term and dynamic development of science curricula and the related implementation procedures. These initiatives were usually accompanied by short courses (summer schools) to introduce the teachers to the new approach and its related scientific background. These courses were usually conducted at science teaching centers located at several academic institutions throughout the country.

In 1992, the *Tomorrow 98* (1992) report on reform in science, technology, and mathematics education in Israel was released. The report included 43 recommendations for special projects, changes, and improvements, both educational and structural, in the area of curriculum development and implementation, and pedagogy of science and mathematics, as well as directions and actions to be taken in the professional development of science and mathematics teachers in general, and the development of leadership among those teachers in particular. More specifically, the report recommended:

- providing science teachers with the opportunity to engage in life-long learning
- creating an environment of collegiality and collaboration among teachers who teach the same or related subjects, which encourages reflection on their work in the classroom
- incorporating the process of change into professional development.

Support for these goals can be found in Loucks-Horsley *et al.* (1998).

To attain these rather demanding goals, national and regional centers for the professional development of science and mathematics teachers were established in Israel (see Hofstein and Even, 2001). The overriding aim of these centers was to enhance educational reform by providing a strong framework for teacher development. These national centers are responsible for, among other activities, the development of science-teacher leaders (in our case chemistry teachers) who are expected to initiate, plan, and implement long-term professional initiatives in their schools, as well as professional development regional centers in their surroundings and nationwide.

9.2.1   Content and Structure of the Chemistry Leadership Program

The chemistry leadership program was based on the assumption that the participating chemistry teachers would be thoughtful learners, that they would be prepared to be professional teacher leaders, and that after completion of the program, they would develop creative strategies for initiating reform in the way chemistry is taught, and in professionalizing other chemistry teachers. Consequently, the program was designed around the following three components:

- developing teachers' understanding of the current trends to include both the content and pedagogy of chemistry learning and teaching. For example, the current trend toward making chemistry more relevant suggests that new programs in chemistry should include, in addition to the conceptual approach and the process of chemistry, societal and personal applications, technological manifestations, and those components that can be characterized as historical and pertaining to the nature of chemistry/science (Kempa, 1983)
- providing the teachers with opportunities to develop personally, professionally, and socially (Bell and Gilbert, 1994)
- developing leadership among these teachers and enhancing their ability to work with other chemistry teachers.

The program was run over a period of two academic years, for a total of 450 hours of professional development activities, conducted one day a week, in an effort to allow for the gradual development and growth of the participants' conceptions, beliefs, and changes in behavior. This was expected to provide enough time for the teachers' personal, professional and social development. The first year of the program was primarily devoted to the development of the teachers' content knowledge (CK) in various chemistry topics characterized as relevant to learners, to provide a historical background for these topics, and to introduce technological ramifications and applications. Topics covered included forensic chemistry, solid-state chemistry, the chemistry of nutrition, and selected topics in the area of interactions between radiation and matter. A large segment of this year was also devoted to the development of the chemistry teachers' pedagogical content knowledge (PCK). The second year was primarily devoted to the development of leadership skills. The various abilities and skills were developed using many of the strategies for professional development suggested by Loucks-Horsley *et al.* (1998) and presented in Figure 9.1. The program for chemistry-teacher leaders was designed to include the necessary components for life-long professional development of science teachers, as well as components that are unique to the development of leadership among chemistry teachers.

Developing

| Content Knowledge | Pedagogical Content Knowledge | Leadership Skills |

Professional Development Strategies

Immersion in the world of science	Curriculum Implementation	Personal Leadership Skills
	Curriculum Replacement Unit	Practicing Leadership Skills
	Immersion into Inquiry	
	Action Research	
	Case Discussions	

Programme Activities

Lectures on	**Workshops on**	**Practice regarding**
Radiation	Dealing with Students' Misconceptions	Developing a Culture of Teamwork
Solid State Chemistry	Varying Teaching Strategies	Tutoring Worskhops
Brain Chemistry	Applying Technological Techniques	Leadership Workshops
Nutrition Chemistry		Planning a Professional Development Programme
Teachign Chemistry by the Inquiry Approach		
Learning and Teaching Theories		

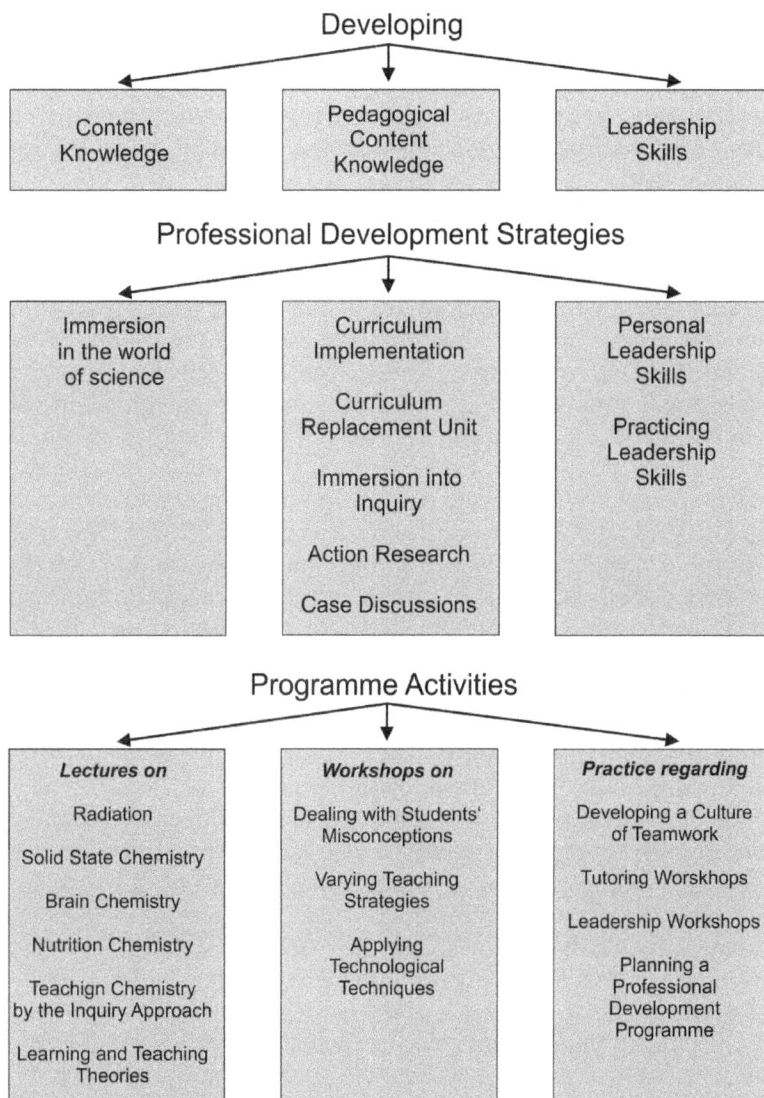

Figure 9.1 Structure of the leadership course.

9.3 Assessment of Teachers' Changes Resulting from the Leadership Program

The assessment of the development of leadership skills in the chemistry teachers focused primarily on the following three interrelated variables:

- development of their personal beliefs about themselves, about teaching chemistry, and about becoming a leader

- development of their professional behavior and activities in their chemistry classroom, focusing primarily on the development of their PCK
- development of leadership skills, and activities involving other chemistry teachers in and outside their schools (teachers' social development).

The teachers who participated in the program were assessed continuously to obtain information regarding these interrelated variables (Hofstein *et al.*, 2003). The assessment was conducted throughout the program and continued for a year after its completion. To increase the validity of the assessment, triangulation was used through a combination of both qualitative (interviews, observations, and reflective protocols on the various meetings) and quantitative (feedback questionnaires, and questionnaires administered among the teachers' students in school) strategies, as well as other tools. The various components of the program assessment provided evidence of the participants' professional, personal, and social growth. This growth could be found in the participants' reports, feedback questionnaires, and interviews that were conducted with a small sample of them throughout the program. Our observations also clearly indicated that the teachers had developed useful social skills and habits. These skills were developed through small-group collaborative discussions and debates on issues regarding students' learning ideas relating to the teaching of chemistry, as well as the professional development of other chemistry teachers (ideas about planning and conducting chemistry workshops and courses). We found that when they entered the program, most of the teachers did not consider themselves as leaders, but rather chemistry teachers who wanted to learn how to become better teachers. Only gradually, through the process of enhancing their CK and through the opportunities provided to develop their personal, professional, and social abilities, did they admit that they were ready to embark on duties that would involve activities requiring leadership. Toward the end of the program, as a result of intensive guidance and involvement in professional development activities, we noted a significant enhancement in the teachers' internalization of the main goals of the leadership program.

These developments would not have occurred if the teachers had not been provided with experiences whose goal was to enhance their chemistry CK and PCK. During the program, the teachers were provided with both numerous and varied opportunities to develop their chemistry knowledge, teaching and assessment skills, and general science education skills. They were also given opportunities to plan and develop learning materials, and develop instructional activities and pedagogical interventions with the goal of varying the classroom learning environment and thus enhancing the students' motivation to learn chemistry. Furthermore, the teachers were provided with opportunities to develop alternative assessment tools that would enable them to implement those tools in their classrooms and in the chemistry classrooms of their peers for whom they were responsible.

Action research was used during the program to provide the teachers with opportunities to assess the impact of the newly developed learning material and pedagogical interventions on their students' learning attitudes and behaviors. Information regarding the teachers' classroom learning environment was obtained by probing their chemistry students' perceptions using paper-and-pencil measures developed within the context of science curriculum development and implementation (Fraser, 1998). In this study, the classroom Learning Environment Inventory (LEI) scale was used (Fraser, 1982). The study revealed a significant change in several dimensions that shape the chemistry classroom learning environment. More specifically changes were revealed in the following dimensions:

- The rate (speed) of the instruction in the classroom was significantly reduced.
- The friction among students in the class was significantly reduced.
- Students' satisfaction with their experiences in the chemistry classroom was significantly increased.
- An increase was observed in the students' perceptions on the LEI scale that assesses goal direction (the extent to which the objectives of learning chemistry are clear).

These changes seemed to result from the experiences provided to the chemistry teachers in the leadership program. Support for the findings regarding changes in students' perceptions of the chemistry classroom learning environment was revealed by the feedback questionnaires gathered from the teachers. In these questionnaires, the teachers reported an increase in their ability to make chemistry more interesting for their students, to cope with students' learning difficulties (by using diagnostic-type tests, for example), and to vary the types of instructional techniques that they adapted for use in their classrooms. The teachers' experiences in the program clearly bolstered their confidence in their ability to try new ideas in the classroom and better plan their activities.

The third component of professional development, based on Bell and Gilbert's (1994) model, is the social component. Social development involves learning to work with others in the educational system in new ways. Our experience suggests that teachers need a strong and solid professional foundation to develop socially. This occurred in the program described here because many of the activities used to enhance the teacher professionally involved the teacher working cooperatively with others in the program and later, in their schools.

Achieving scientific literacy for all, as described in Chapter 6 of this book, has become a national goal for education in many countries. Although admirable, this goal represents a challenge for science teachers and for those responsible for professional development. Achieving this goal must be accompanied by a reform in the way chemistry is taught in schools. This is based on the assumption that teachers are the best professional partners to develop lesson plans that combine issues related to chemistry CK and its associated PCK. Clearly, teachers know their students' needs, interests, and abilities best. Those who become leading teachers have the potential, when working with other teachers, to initiate and develop chemistry teachers' communities of practice.

To sum up, it is clear that the teachers are at the center of the network of influence, thereby avoiding or reducing "top-down" curricular procedures. Teachers need time to develop as policy-makers and to effect changes.

Professional development programs conducted in the 21st century need to build on lessons learned from the "golden age" of science curriculum development and implementation and the first generation of the National Standards of Science Education (NRC, 1996). History has taught us that to make the learning of science more relevant to students, we need to involve teachers – and especially leading teachers – in discussions of issues related to the content and pedagogy of the teaching and learning. This approach was applied to the development of learning materials in the UK in the Salters project (Bennet and Lubben, 2006) and more recently, in the USA in the design of the NGSS. This approach has also been used throughout the process of developing additions to the ACS *Chemistry in the Community* textbook, now in its 6th edition. While in the 1960s, teachers' professional development procedures were conducted on the basis of "top-down" initiatives, in which most of the experiences were dictated by academic institutions and or governments, at the beginning of the 21st century, we are observing a more "bottom-up" approach, in which many of the decisions made regarding content and pedagogy of teaching chemistry involve teachers who have undergone long-term, leadership-type professional experiences (see for example in the USA, Project 2061; AAAS, 1993).

9.4 The Content, Structure, and Activities of the Chemistry Teachers' Leadership Development Program Conducted at the Weizmann Institute of Science in Israel

In their book, Loucks-Horsley *et al.* (1998, p. 201) posed the question: "What specific roles of teacher leader are we interested in developing?" These are the three main abilities:

- **Teacher development**: we expect leading teachers to be involved in initiating, facilitating, planning, and conducting professional development initiatives for chemistry teachers in their schools and/or in their region.
- **Teachers as curriculum developers**: we expect participants in professional development of leading chemistry teachers to be able to develop chemistry curricula, innovative instructional and pedagogical techniques, and new assessment approaches aligned with these pedagogies.

- **Teachers as active participants:** participants should be able to create an effective, school-based learning environment that will involve communities of practice and networking to improve school-based pedagogy. They should also be able to support new teachers in their initial years of practice in the chemistry classroom.

9.5 Summary

- The main goal of the program presented in this chapter was the long-term development of teacher leaders who would support and help attain goals of reform (in this case the reform taking place in Israel). The reform in chemistry education in Israel addresses both the content of chemistry and the pedagogy of chemistry, namely in the instructional techniques and learning methods implemented in the chemistry classroom to make it a more educationally effective learning environment.
- The model that was adopted for this program was originally developed by Bell and Gilbert (1994) in New Zealand. They suggested that science teacher development be viewed as *professional*, *social*, and *personal* development and that teacher development programs and activities should address these three interrelated components.
- The professional development program detailed in this chapter was developed with the goal of obtaining changes in these three aspects. The results of the assessment of the teachers' development throughout the program provided evidence of the effectiveness of the experiences and content provided to them through the various professional development strategies used. These aimed at enhancing the teachers' CK, PCK and leadership skills.
- Regarding teachers' development at the personal level, we presented evidence (from both quantitative and qualitative sources) that as a result of their experiences, the teachers developed effectively. This development involved attending to feelings about the change process that they underwent, about the changes they underwent as chemistry teachers, and about the confidence that they gained (over time) regarding the idea that they might become leaders in chemistry education.
- Professional development relates mainly to teachers' development in the content of the subject matter that they teach and in the relevant PCK. Evidence for this component was gathered from students' perceptions of the chemistry classroom learning environment, as well as from the teachers' self-reports regarding the changes that they underwent, which they applied in their classroom practice in their own schools, and in activities outside of the school, namely in science teachers' professional development centers.
- The teachers had many opportunities to enhance their social skills through collaborations and cooperation with their peers in the program, through working with the team of chemistry teachers in their

own schools and at a later stage, through professional development activities as tutors in professional development programs.

- Until the 1990s, effort was mostly invested in trying to achieve the desired changes in school science that focused on the development of improved science curricula. In the last decade, however, more attention has been gradually paid to the teacher, because past efforts in educational reform suggested that the teacher plays a critical role in the ways new ideas are created in the classroom.
- The establishment of regional teacher centers has created a comprehensive framework that can provide opportunities for in-service chemistry teachers to achieve life-long learning in their profession.

References

AAAS, (1993), *Benchmarks for Scientific Literacy*, Washington: AAAS.

Bennet J. and Lubben F., (2006), Context-based chemistry: the Salters approach, *Int. J. Sci. Educ.*, **28**, 999–1015.

Bell B. and Gilbert J., (1994), Teacher development as personal, professional, and social development, *Teaching and Teacher Education*, **10**, 483–497.

Bodner G. M., (2011), Preparing chemistry teachers for the next generation science standards, *Chem. Eng. News*, **89**(50), 32.

Bybee R. W., (1993), *Reforming Science Education*, New York: Teachers College Press.

Fraser B. J., (1982), *Assessment of Learning Environments: Manual for Learning Environment Inventory (LEI) and My Class Inventory (MCI), 3rd version*, Perth, Western Australia: Curtin University.

Fraser B. J., (1998), Classroom environment instruments: development, validity, and application, *Learn. Environ. Res.*, **1**, 7–33.

Friel S. N. and Bright G. W., (1997), *Reflecting on Our Work: N.S.F. Teacher Enhancement in K-6 Mathematics*, Lanham: University Press of America.

Fullan M. G., (1991), *New Meaning of Educational Change*, New York: Teachers College Press.

Hofstein A., Carmi M. and Ben-Zvi R., (2003), The development of leadership among chemistry teachers in Israel, *Int. J. Sci. Math. Educ.*, **1**(1), 39–65.

Hofstein A. and Even R., (2001), Developing chemistry and mathematics teacher-leaders in Israel, in Nesbit C. R., Wallace J. D., Pugalee D. K., Courtny-Miller A. and DiBiase W. J., (ed.), *Developing Teacher-leaders*, Columbus: ERIC Clearing House, pp. 189–208.

Kempa R. F., (1983), Developing new perspectives in chemical education, *Proceedings of the 7th International Conference in Chemistry, Education, and Society*, Montpellier, France, pp. 34–42.

Lawrenz F., (2001), Evaluation of teacher leader professional development, in Nesbit C. R., Wallace J. D., Pugalee D. K., Courtny-Miller A. and DiBiase W. J., (ed.), *Developing Teacher Leaders*, Columbus: ERIC Clearing House.

Loucks-Horsley S., Hewson P. W., Love N. and Stiles K. E., (1998), *Designing Professional Development for Teachers of Science and Mathematics*, Thousand Oaks: Corwin.

National Research Council (NRC), (1996), *National Science Education Standards*, Washington: National Academies Press.

NGSS Lead States, 2013, *Next Generation Science Standards: For States, by States*, Washington, DC: National Academies Press.

Pratt H., (2001), The role of the science leader in implementing standard-based science programs, in Rohton J. and Bowers P. (ed.), *Professional Development, Leadership, and the Diverse Learner*, Washington: NSTA, pp. 1–10.

CHAPTER 10

The Professional Development of Chemistry Teachers – A Summary

The authors of the current book have tried to convince the readers that there can be many different elements of professional development efforts for chemistry teachers. At times, the goal is to help teachers bring contemporary knowledge and ideas from the frontiers of the chemical enterprise – such as nanoscale science – into their classrooms. Or, as we have seen, it might involve bringing real-world examples into chemistry courses. It can focus on improving teachers' understanding of the nature of science. It can focus on learning both in and by the chemistry laboratory, with appropriate attention being paid to safety and "prudent practice". At times, the goal is to broaden the student population that takes chemistry, moving toward a "chemistry for all" perspective. This might involve incorporating socio-scientific issues (SSIs), or the use of modern information and communication technologies (ICT) in chemistry classes. Regardless of the content of professional development efforts, the authors contend that the teacher is the central key and driver for implementing any reform or effective practice in the chemistry classroom (Hattie, 2008). With this view in mind, the final chapter of this book summarizes and reflects on its key issues.

10.1 A Summary and Outlook

This book has provided insights into models for effective professional development of chemistry teachers that share the common feature of answering the call for "active learning" by students (Bonwell and Eison, 1991). For more than 30 years, proponents of the constructivist theory of knowledge

Advances in Chemistry Education Series No. 1
Professional Development of Chemistry Teachers: Theory and Practice
By Rachel Mamlok-Naaman, Ingo Eilks, George Bodner and Avi Hofstein
© Rachel Mamlok-Naaman, Ingo Eilks, George Bodner and Avi Hofstein 2022
Published by the Royal Society of Chemistry, www.rsc.org

have argued that "knowledge is constructed in the mind of the learner" (Bodner, 1986). This book suggests that this is equally true for students and their instructors. In much the same way that active learning is valued more than passive approaches in the classroom environment, programs whose goal is to shape what happens in these classrooms cannot assume that the participants in professional development efforts are passive recipients of information about either content or its related pedagogy. For professional development programs to succeed in creating long-lasting changes, the teachers enrolled in these programs must be active participants in the workshops.

Earlier in this book, the process for the development of the American Association of Chemistry Teachers (AACT) was described (Bodner, 2013). Those of us involved in this process noted that the parent organization, the American Chemical Society (ACS), assumed that teachers were best served when the society did things *to* chemistry teachers. With time, a more enlightened perspective was introduced in which the ACS did things *for* chemistry teachers. The AACT reflects our present understanding of curriculum and pedagogy reform; successful programs only occur when things are done *with* chemistry teachers.

Each of this book's authors came, in their own way, to the same conclusion on the basis of research into the professional development environment. Stand-alone professional development programs do not give rise to long-lasting improvements in either curriculum or pedagogy. Although there are substantial differences between the educational systems in Israel, Germany, and the USA, we all recognize that efforts to bring about long-lasting change need to keep the teachers together, learning from each other, and being accompanied – at least in their first steps of change implementation – by external experts. In many ways, this group of teachers should become a "community of practice" (Wenger, 1998, 2000).

Different terminologies are used in different countries to express the same idea. In the USA, the most common way of describing the approach advocated in this book is *continuing professional development*. Teachers are expected to earn "continuing education units". In other countries and in European Union educational policy, it is called *continuous professional development*. Regardless of the term used to describe the process, successful professional development efforts do not stop at the end of a one-day or a five-day workshop.

The theories of teaching and learning, educational research, and experiences gained as a result of successful implementation of professional development all suggest that continuing professional development of chemistry teachers needs to address all three professional knowledge domains of teachers described by Shulman (1986). Teachers need to refresh their content knowledge (CK) to keep up with the constantly developing body of knowledge in chemistry and its related applications; they have to be familiarized with current developments in general pedagogical knowledge (PK) about teaching and learning, and develop their chemistry-specific pedagogical content knowledge (PCK). In all three of these domains, knowledge is developing so rapidly that individual teacher knowledge becomes outdated in the absence of continuous renewal. The process of renewal or refreshment of teachers' knowledge, however, needs to take into account existing knowledge and associated attitudes and beliefs formed during pre-service teacher education, as well as the experiences the teachers have had while working as chemistry teachers (Van Driel *et al.*, 1998).

Continuing professional development of chemistry teachers should start with the teachers' prior knowledge and should take all four domains of teachers' professional growth into account, as described in the Interconnected Model of Teacher Professional Growth (IMTPG) (Clarke and Hollingsworth, 2002). These domains are: the personal domain (beliefs, attitudes, and previous experiences), the practical domain (the authentic teaching practices of the teacher), the external domain (topic requirements, media, and curriculum aspects), and the domain of consequences (goals and effects). As we have seen, the professional development of chemistry teachers should involve a long-term commitment, inspired and supported by external experts. It should include phases of active involvement in the planning, implementation, and structuring of chemistry lessons. A combination of all of these processes can help implement changes in the classroom effectively, leading to positive effects in teaching and learning (Mamlok-Naaman *et al.*, 2005).

Traditional models of chemistry teacher professional development are generally based on a "top-down" model of dissemination of content and its related instructional techniques. They provide teachers with the results of both general and domain-specific educational research, innovative teaching ideas, and new teaching and learning materials. Media channels cover print and online publications, teacher conferences, and half- or full-day workshops (Mamlok-Naaman *et al.*, 2013). In the absence of better options for chemistry teacher professional development, such as those introduced in this book, these traditional activities remain prominent. Today, however, we face an ever-changing media landscape. It is not yet clear how the continuing developments in digital technologies and social media will change the nature and possibilities of teacher professional development, and what effects this change will have. More communication among teachers and between teachers and professional development providers might occur as a result of new communication technologies. However, an overflow of information and decreasing quality control might be another result. What is clear is that traditional means of teacher professional development need to be rethought in light of the new and emerging technologies in digital and social media. As stated by Van Driel and de Jong (2015), "there is a lack of empirical research on the role and impact of new technologies and online professional networks on chemistry teachers' knowledge, beliefs, and classroom activities" (p. 117).

Ideally, professional development of chemistry teachers should involve long-term strategies to allow teachers to become familiar with new information and develop ownership. One-way dissemination of information and single-contact events have limited potential. Potentially better results can be achieved by repeated face-to-face contact with other teachers who are participating in the same professional development program and with external experts from outside the teacher's own school (Richardson, 2003). Clearly, any professional development should take into account the prior knowledge, experiences, and beliefs of the participants and connect them to the new content and teaching ideas (Van Driel *et al.*, 1998). It should relate the new content or teaching strategies to teachers' practical experience and reflections when introducing more advanced teaching strategies. The examples reported in the various chapters of this book can serve as patterns for short- and long-term, top-down and bottom-up professional development of chemistry teachers because research about them has provided many promising results.

Better approaches to professional development based on "bottom-up" models or mixed strategies, which include top-down input but also actively involve teachers in the development of innovative practices, have the potential to achieve more powerful and long-lasting results (Loucks-Horsley *et al.*, 1998). One of the examples reported in Chapter 4 of this book refers to encouraging teachers to develop ownership of a curriculum innovation by actively involving them in its adoption, development, and implementation. Turning teachers into curriculum developers and innovators of their own practice is one of the most promising ways for both teacher professional growth and the effective implementation of curriculum change. Moreover, it encourages the creation and development of professional learning communities (PLC), which are an effective, bottom-up way of bringing innovation into the chemistry curriculum. A major topic for discussion in PLC workshops for chemistry teachers should be the diagnosis of students' ideas, difficulties, and misconceptions, with the aim of matching new teaching strategies to the students' individual learning characteristics (Westwood, 2001). The PLC work, however, should also include joint reflections about the teachers' own ideas, difficulties, and conceptions to identify and overcome lacks in the teachers' knowledge. Another focus should be the growing heterogeneity and diversity in many chemistry classrooms worldwide, with the related challenges for science education and science education research (Markic *et al.*, 2012). Students with different abilities, interests, and levels of motivation should be offered differentiated instruction that meets their individual needs (Bell and Linn, 2002), moving toward personalized teaching and learning approaches (Hodson, 1998).

Another bottom-up model of professional development that is discussed extensively in Chapter 5 of this book is action research. The goal of action research is to make every teacher a researcher within the context of their own classroom. Action research develops concrete teaching practices in a cyclical approach of identifying deficits, implementing changes, and studying and understanding corresponding effects for further improvement (Laudonia *et al.*, 2017). In chemistry education, teacher professional development through action research can be achieved using a range of models, from accompanying teachers in individual chemistry education action research case studies to involving them in long-term collaborative projects of curriculum and practice improvement (Mamlok-Naaman and Eilks, 2012). The authors believe that action research is one of the most emancipatory strategies in teacher professional growth because it enables the teachers to understand and implement evidence-based changes in their classrooms on their own. Action research is being advocated more and more often in educational policy documents, such as the UNESCO report by Alidou and Glanz (2015). It is recommended as a promising strategy for teachers' lifelong learning and development of innovations in educational practices. The literature on its implementation in science education is growing. However, further investment in the broad implementation of action research is needed (Laudonia *et al.*, 2017). Improvements should also involve making better use of the solutions developed by teachers through action research, because most of the teachers' findings never get published. Solutions are found, but are often not shared among teachers.

Besides providing models of professional development, this book also describes fields for which professional development is particularly relevant. Chemistry teachers repeatedly encounter challenges in pursuing the goals of teaching and learning chemistry due to changes in educational policy. At one time, the goal of chemistry education in schools was to identify, motivate, and properly prepare those students who would later embark on careers in science, medicine, or engineering. In the 1980s, educational policy began to shift toward the assumption that learning science is an important goal for every future citizen, and therefore every student. Approaches changed to curricula that were more oriented toward everyday life, society, or technological contexts (Eilks *et al.*, 2013). Today, curricula are being recommended that focus on controversial SSIs and the interplay between chemistry, technology, society, and the environment (Eilks *et al.*, 2013). A related topic, Education for Sustainable Development (ESD), has also emerged as another highly relevant educational paradigm. The United Nations suggested including ESD in all educational domains and levels, including chemistry (UNCED, 1992). The move toward including SSIs and ESD in the chemistry curriculum is part of a reshaping of the goals for all science education toward a more critical vision of scientific literacy (Sjöström *et al.*, 2017). Because of the importance of chemistry for the sustainable development of our future, chemistry education has a special responsibility in contributing to ESD (Burmeister *et al.*, 2012). The SSIs related to sustainable development, however, are complex and generally of an interdisciplinary nature, *e.g.*, how to react to challenges caused by climate change or by limitations in critical natural resources. SSIs and society's beliefs about them change over time, and newly developed knowledge can provide a different perspective. As a result, teachers need to keep their chemistry knowledge continuously updated. These topics, however, also require teachers to develop interdisciplinary knowledge and, at the same time, learn how to deal with multidisciplinarity and uncertainty. Teachers who teach SSIs have to learn how to handle discussions, debate and uncertainty in the chemistry classroom and they have to acquire a new portfolio of instructional techniques, such as how to guide role-playing or how to structure debates (Sjöström *et al.*, 2015). Thomas (2010) claimed that education for sustainability requires changes in both curricula and transformative pedagogy, focusing on the processes of learning rather than the accumulation of knowledge, in order to enable students to be innovative, creative, and critical. Gaining skills such as interdisciplinary thinking, problem solving, working in teams, and holistic thinking are therefore critical in education related to SSIs and sustainability.

Another important issue for chemistry professional development involves learning how to operate laboratory/practical work more effectively. It is generally agreed that a particularly important component of chemistry education is the laboratory. The science education literature suggests that the science (in our current case chemistry) laboratory is a unique environment with respect to both teaching and learning. The chemistry laboratory has a long tradition as a practice of teaching and learning chemistry (Childs and Flaherty, 2016). There is, however, general agreement among researchers who have focused on the chemistry laboratory that it is not being used to its full potential for learning (Hofstein and Lunetta, 2004). In the past, in teaching and learning, the chemistry laboratory focused far too often on manipulating apparatus and chemicals, instead of on "manipulating" ideas and thought. Today, the implementation of new, more effective practices based on research is being suggested to develop a broader spectrum of skills in the chemistry laboratory, especially inquiry and problem-solving skills. Appropriate pedagogy in the chemistry laboratory to promote this new set of skills is very demanding for the chemistry teacher in terms of guiding the students and attaining goals of inquiry and other higher-order learning skills. Models for professional development concerning practical work should be highly aligned with the goals and skills that we expect the students to acquire. To attain these goals, the chemistry education community needs to develop effective professional development procedures that will help teachers experience the inquiry process on its different levels, understand how it works and develop a firm belief in its success. Non-formal educational practices in out-of-school laboratory learning environments, which have emerged in recent years in many countries, are a good place to innovate practical work in chemistry education, to enable students to practice it, and to allow teachers to learn about it (Garner *et al.*, 2015). If we want the chemistry laboratory to function as an effective learning platform, we need to provide the teachers with some kind of toolbox to act effectively in such practical activities, as well as with professional development on how to use it.

The most rapidly changing challenge for contemporary teacher professional development is probably today's constantly changing digital media world. The changes in digital and social media create challenges for teacher professional development because they alter public and student access to, and representation of, chemistry-related information. Used in the right way, modern ICT offer unique opportunities to support the teaching and learning of chemistry (Dori *et al.*, 2013). ICT can help to better integrate chemistry learning by incorporating its different representations, namely the macroscopic, submicroscopic, symbolic, and contextual representational levels, although information overload should be avoided. Teachers must familiarize themselves with these changes, both from a technical perspective and in terms of how chemistry-related information is represented in the new media. They need to invest in both their skills in operating current technologies and improving their knowledge about how to use ICT to support chemistry learning. They need to develop contemporary, chemistry-specific technological pedagogical content knowledge (TPACK) (Mishra and Koehler, 2006). Teachers also need to learn how to cope with the fact that a growing proportion of the young generation, the "digital natives", will in some respects be more skillful at operating modern ICT than many of their teachers. This situation will, however, also offer chances to more actively involve the students in the teaching and learning processes in chemistry education, *e.g.*, by asking students to develop their own animations, illustrations and learning tools using modern ICT (Krause and Eilks, 2017). The technological changes in ICT are occurring so rapidly that educational research cannot keep up. Educational studies take a lot of time, and they cannot always determine how a certain technology will affect learning before the corresponding technology is replaced by a newer one. Thus professional development in the field of how to use ICT in chemistry teaching might be based more on practices that work well as opposed to traditional educational research.

Finally, the current book focuses on teacher professional development to form leading teachers, who will eventually support educational reforms related to the various components of the teaching and learning of chemistry. A spectrum is needed among organizers of curriculum development and implementation and the practicing teachers, with leading teachers linking these two ends of the continuum to both enable and mediate educational reform implementation. These leading teachers need to be experts in both the content of chemistry and the instructional techniques and learning methods to be implemented in the chemistry classroom. They should also be experts in dealing with adult learning and, hopefully, skillful in reflecting on and collecting feedback from the teachers about both positive effects and problems that need to be reported to the reform initiators. The model proposed by Bell and Gilbert (1994) suggests that science teacher professional development should be viewed as professional, social, and personal development, and that teacher development programs and activities should address these three interrelated components. Teachers can enhance their social and personal skills through collaborations and cooperation with peers, through working within the team of chemistry teachers in their own schools, and at a later stage, in activities as tutors in professional development programs. A focus on social and personal skills should be a component of professional development designed to promote prospective leading teachers, because they need leadership skills and have to be sensitive when communicating, guiding and supervising both experienced and less experienced colleagues.

10.2   Final Remarks

This book reports on theoretical views and practical examples of chemistry teacher professional development from many national and international projects with which the authors have been associated. It reflects decades of experiences in chemistry teacher pre-service and in-service professional development in Israel, Germany, and the USA. The authors hope that the views and examples discussed in this book will inform providers of chemistry teacher professional development in various countries and give them ideas to further develop their own practices for the sake of our students' more effective learning. It needs to be said, however, that the professional development of chemistry teachers is a field that needs more research. In particular, "best practices" for creating both an effective structure and implementation of professional development courses and programs need to be determined. Professional development would further benefit from more empirical research analyzing the preconditions, progressions and effects of chemistry teacher professional development, and more design research to develop further models, strategies, and materials. It is the hope of the authors that the examples described in this book will contribute to this endeavor.

References

Alidou H. and Glanz C., (2015), *Action Research to Improve Youth and Adult Literacy*, Abuja: UIL.

Bell B. and Gilbert J., (1994), Teacher development as personal, professional, and social development, *Teaching and Teacher Education*, **10**, 483–497.

Bell P. and Linn M. C., (2002), Beliefs about science: how does science instruction contribute? in Hofer B. K. and Pintrich P. R. (ed.), *Personal Epistemology: The Psychology of Beliefs about Knowledge and Knowing*, Mahwah: Lawrence Erlbaum, pp. 321–346.

Bodner G. M., (1986), Constructivism: a theory of knowledge, *J. Chem. Educ.*, **63**, 873.

Bodner G. M., (2013), Creation of an American Association of Chemistry Teachers, *J. Chem. Educ.*, **91**, 3–5.

Bonwell C. C. and Eison J. A., (1991), Active learning: creating excitement in the classroom, *1991 ASHE-ERIC Higher Education Reports*, Washington: ERIC Clearinghouse on Higher Education.

Burmeister M., Rauch F. and Eilks I., (2012), Education for Sustainable Development (ESD) and secondary chemistry education, *Chem. Educ. Res. Pract.*, **13**, 59–68.

Childs P. E. and Flaherty A., (2016), From elaboratory to e-laboratory: on the history of practical work in school science, in Eilks I., Markic S. and Ralle B. (ed.), *Science Education Research and Practical Work*, Aachen: Shaker, pp. 13–26.

Clarke D. and Hollingsworth H., (2002), Elaborating a model of teacher professional growth, *Teaching and Teacher Education*, **18**, 947–967.

Dori Y. J., Schanze S. and Rodrigues S., (2013), How to promote chemistry learning through the use of ICT, in Eilks I. and Hofstein A. (ed.), *Teaching Chemistry – A Studybook*, Rotterdam: Sense, pp. 213–241.

Eilks I., Ralle B., Rauch F. and Hofstein A., (2013), How to balance the chemistry curriculum between science and society, in Eilks I. and Hofstein A. (ed.), *Teaching Chemistry – A Studybook*, Rotterdam: Sense, pp. 1–36.

Garner N., Siol A. and Eilks I., (2015), The potential of non-formal laboratory environments for innovating the chemistry curriculum and promoting secondary school level students education for sustainability, *Sustainability*, 7, 1798–1818.

Hattie J., (2008), *Visible Learning*, Routledge, London.

Hodson D., (1998), *Teaching and Learning Science: Towards a Personalized Approach*, Buckingham: Open University Press.

Hofstein A. and Lunetta V. N., (2004), The laboratory in science education: foundations for the twenty-first century, *Sci. Educ.*, **88**, 28–54.

Krause M. and Eilks I., (2017), Learning about the nomenclature of organic substances by the creation of stop-motion videos, *Chem. Act.*, **109**, 36–38.

Laudonia I., Mamlok-Naaman R., Abels S. and Eilks I., (2017), Action research in science education – an analytical review of the literature, *Educ. Action Res.*, DOI: 10.1080/09650792.2017.1358198., advance article

Loucks-Horsley S., Hewson P. W., Love N. and Stiles K. E., (1998), *Designing Professional Development for Teachers of Science and Mathematics*, Thousand Oaks: Corwin Press.

Mamlok-Naaman R., Ben-Zvi R., Hofstein A., Menis J. and Erduran S., (2005), Influencing students' attitudes towards science by exposing them to a historical approach, *Int. J. Sci. Math. Educ.*, **3**, 485–507.

Mamlok-Naaman R. and Eilks I., (2012), Different types of action research to promote chemistry teachers' professional development – a joined theoretical reflection on two cases from Israel and Germany, *Int. J. Sci. Math. Educ.*, **10**, 581–610.

Mamlok-Naaman R., Rauch F., Markic S. and Fernandez C., (2013), How to keep myself being a professional chemistry teacher, in Eilks I. and Hofstein A. (ed.), *Teaching Chemistry – A Studybook*, Rotterdam: Sense, pp. 1–36.

Markic S., Eilks I., di Fuccia D. and Ralle B., (2012), *Heterogeneity and Cultural Diversity in Science Education and Science Education Research*, Aachen: Shaker.

Mishra P. and Koehler M. J., (2006), Technological pedagogical content knowledge: a framework for teacher knowledge, *Teach. Coll. Rec.*, **108**, 1017–1054.

Richardson V., (2003), The dilemmas of professional development, *Phi Delta Kappa*, **84**(5), 401–406.

Shulman L. S., (1986), Those who understand: knowledge growth in teaching, *Educ. Res.*, **15**(2), 4–14.

Sjöström J., Frerichs N., Zuin V. G. and Eilks I., (2017), The use of the concept of Bildung in the international literature in science education and its implications for the teaching and learning of science, *Stud. Sci. Educ.*, **53**, 165–192.

Sjöström J., Rauch F., Eilks I., (2015), Chemistry education for sustainability, in Eilks I. and Hofstein A. (ed.), *Relevant Chemistry Education – From Theory to Practice*, Rotterdam: Sense, pp. 163–184.

Thomas I., (2010), Critical thinking, transformative learning, sustainable education, and problem-based learning in universities, *J. Transformative Educ.*, 7, 245–264.

UNCED, (1992), *Agenda 21*, retrieved from the World Wide Web, July 10, 2011 at http://www.un.org/esa/dsd/agenda21/.

Van Driel J. H., Verloop N. and de Vos W., (1998), Developing science teachers' pedagogical content knowledge, *J. Res. Sci. Teach.*, 35, 673–695.

Van Driel J. H. and de Jong O., (2015), Empowering chemistry teachers' learning: practices and new challenge, in Garcia-Martinez J. and Serrano-Torregrosa E. (ed.), *Chemistry Education Best Practices, Opportunities and Trends*, Germany: Wiley-VCH, pp. 114–118.

Wenger E., (1998), *Communities of Practice: Learning, Meaning, and Identity*, Cambridge: Cambridge University Press.

Wenger E., (2000), Communities of practice and social learning systems, *Organization*, 7, 225–246.

Westwood P., (2001), Differentiation' as a strategy for inclusive classroom practice: some difficulties identified, *Aust. J. Learn. Disabilities*, **6**, 5–11.

Subject Index

References to figures are given in *italic* type. References to tables are given in **bold** type.